Digital Humanism

Marta Bertolaso · Luca Capone ·
Carlos Rodríguez-Lluesma
Editors

Digital Humanism

A Human-Centric Approach to Digital
Technologies

Editors
Marta Bertolaso 🆔
Campus Bio-Medico University
of Rome
Rome, Italy

Luca Capone 🆔
Campus Bio-Medico University
of Rome
Rome, Italy

Carlos Rodríguez-Lluesma 🆔
IESE Business School-University
of Navarra
Madrid, Spain

ISBN 978-3-030-97053-6   ISBN 978-3-030-97054-3   (eBook)
https://doi.org/10.1007/978-3-030-97054-3

This Palgrave Macmillan imprint is published by the registered company Springer Nature Switzerland AG
The registered company address is: Gewerbestrasse 11, 6330 Cham, Switzerland

Foreword

The Social Trends Institute (STI) is a non-profit international research centre dedicated to fostering understanding of globally significant social trends. Founded in New York City, STI also has a delegation in Barcelona, Spain. The president of STI is Carlos Cavallé, Ph.D.

The people and institutions that support STI share a conception of society and the individual that commands a deep respect for the equal dignity of human beings and for freedom of thought, as well as a strong desire to contribute to social progress and the common good. To this end, STI organizes Experts' Meetings around specific topics in its areas of priority study. These meetings take an interdisciplinary and international approach, bringing together the world's leading thinkers. Currently, STI prioritizes research that advances knowledge and practice in three areas: "The Future of Work", "Technology and Ethics" and "The Thriving Society". Findings are disseminated to the media and through scholarly publications.

This volume, *Digital Humanism: A Human-Centric Approach to Digital Technologies*, is the result of a series of Experts' Meetings held online in April and May 2021, under the academic leadership of Marta Bertolaso, Luca Capone and Carlos Rodríguez-Lluesma. The series explored the concept of "digital humanism" from a multidisciplinary point of view: a particular focus was the involvement of social sciences and the humanities as supporting scientific knowledge and practices related to digital technologies.

The results of the research carried out under the auspices of these Experts' Meetings are presented in this book. Without endorsing any particular viewpoint, STI hopes that as a whole, the contributions collected in this volume will deepen readers' understanding of the concept of "digital humanism".

2022

Tracey O'Donnell
Secretary General
Social Trends Institute
Barcelona, Spain

ACKNOWLEDGEMENTS

We would like to thank the Social Trends Institute (www.socialtrendsins titute.org) for its generous financial and organizational support, which enabled the Experts' Meeting series *Digital Humanism: A Human-Centric Approach to Digital Technologies* (Campus Bio-Medico University of Rome, April 9, May 7, May 28, 2021). The present volume is based on the papers collected and extensively discussed during these meetings.

Marta Bertolaso
Luca Capone
Carlos Rodríguez-Lluesma

CONTENTS

Notes on Contributors

Javed Ali is a research scholar and Ph.D. candidate at the Indian Institute of Technology, Kharagpur. He is a visiting AI researcher at deppakpha.com. Javed's expertise and interest lie in machine learning and deep learning, as well as geophysics. He is currently pursuing his dissertation in carbon-capture storage and utilization technologies.

Marta Bertolaso is Professor of the Philosophy of Science and Human Development at the University Campus Bio-Medico of Rome and Adjunct Professor at the University of Bergen. She works on epistemological issues involved in understanding complex systems with a primary focus on living systems and developments in new artificial intelligence technologies.

Aninda Bhattacharjee is a big-data engineer with three years' experience in the field of innovative technologies. His interests are NLP, and emerging technologies to make deploying and scaling machine-learning applications easy and thus production ready. In his spare time he experiments with various graph-based algorithms and architectures.

Luca Capone is a Ph.D. student at the University Campus Bio-Medico of Rome and a Fellow of the Philosophy of Science and Human Development research unit. He is currently working on artificial intelligence models applied to natural language.

Mario Di Giulio is partner responsible for the FinTech Focus Group at law firm Pavia e Ansaldo, and co-founder and vice president at The Thinking Watermill Society, a non-profit organization that promotes debates, inter alia, on the use of the new technologies and their impact on society.

Marta M. Elvira is Professor in the Departments of Strategic Management and Managing People in Organizations at IESE Business School, Spain. She earned her Ph.D. in Organizational Behaviour and Industrial Relations at University of California, Berkeley. Prior to joining IESE, she was Academic Dean at Lexington College, Chicago, tenured Associate Professor at the University of California, Irvine, and Associate Professor of Organizational Behaviour at INSEAD, France. She has been a visiting scholar at MIT's Sloan School of Management, the Institute for Work and Employment Research and the Instituto Tecnológico de Monterrey, Mexico. Marta's work examines the political and economic processes involved in designing organizational reward structures, and their resulting effects on employee earnings and careers. She has co-edited three books and her articles have been published in *Academy of Management Journal*, *Organization Science*, *Work and Occupations*, *Industrial Relations* and *ILR Review*, among other journals.

Mei Lin Fung Chair and co-founder in 2015 of the People-Centered Internet, is a technology pioneer working to ensure that technology works for humanity as the next 3.9 billion people come online. She serves as Vice-chair for Internet Inclusion at the Institute of Electrical and Electronic Engineers' (IEEE) Internet Initiative and Smart Village. She is on the World Economic Forum (WEF) Steering Committee for Internet for AI and facilitated the Working Group on Cross-Border Data Flows at the WEF ASEAN Summit in Hanoi in 2018. Mei Lin was a 2018 Woman of the Year finalist at the Silicon Valley Women in IT Awards organized by Information Age.

Zora Kovacic is a researcher at the Universitat Oberta de Catalunya and Associate Professor II at the Centre for the Study of the Sciences and the Humanities (SVT), University of Bergen. Zora holds a B.A. in economics and development studies from the School of Oriental and African Studies, London, an M.Sc. in environmental studies and a Ph.D. in environmental science and technology from the Institute of Environmental Science and

Technology (ICTA), Autonomous University of Barcelona. She investigates how scientific information and innovation are used in policy-making when dealing with complex environmental challenges characterized by uncertainty and pluralism.

Migle Laukyte earned her Ph.D. at the Bologna University School of Law, Italy and spent the following two years as Max Weber Postdoctoral Fellow at the European University Institute (EUI), Italy. She is currently a tenure track professor in cyberlaw and cyber rights at the University Pompeu Fabra in Barcelona, Spain. She was a visiting professor at the Bartolomé de las Casas Human Rights Institute at the Universidad Carlos III de Madrid, Spain in 2019 and CONEX-Marie Curie Fellow, working on the ALLIES (Artificially InteLLigent EntitIES: Their Legal Status in the Future) project which was dedicated to elaborating a draft model of legal personhood for AI. Her research interests are legal, ethical and philosophical questions related to AI, robotics and other disruptive technologies.

Leng Leroy Lim is an executive coach and change management consultant who specializes in working with leaders who are ready to build a legacy, helping them to make wise choices and to lead with a powerful vision. Leng coaches on a broad range of topics from teamwork to leadership succession to merging strategic vision/execution with legacy and personal power.

Tannistha Maiti holds a bachelor's degree from IIT Kharagpur, India, an MSc from Virginia Tech and a Ph.D. in geophysics and seismology from the University of Calgary. She has worked on various topics ranging from earth science to computation and mathematical modelling, and is particularly interested in machine-learning applications in the energy and healthcare sectors. Tannistha has more than 20 peer-reviewed publications in various conferences and journals. She leads ML OPs engineering practice within deepkapha AI Lab, where she supervises several deep-learning internships.

Pierluigi Malavasi is Professor of General and Social Pedagogy and Director of the Master's Courses in Management and Communication of Sustainability, and Training, Green Jobs and the Circular Economy at the Catholic University of the Sacred Heart, Rome.

Antonio Malo is Professor of Anthropology at the Pontifical University of the Holy Cross. Director of the Editorial Board of the Faculty of Philosophy at the same university, he is a member of the Scientific Committee of the Centro para el Estudio de las Relaciones Interpersonales (CERI) of the Austral University, Argentina, founding member of the Research Centre for Relational Ontology (ROR) of the Pontifical University of the Holy Cross and a visiting professor in various American universities. His writings and research interests include the anthropology of affectivity, action theory and human relationships.

Stefano Marrone is a researcher at the University of Naples Federico II. His research topics are within the area of pattern recognition and computer vision, with applications ranging from biomedical image processing to remote sensing and image/video forensics. More recently, he has also been working on ethics, fairness and privacy in artificial intelligence, spending eight months at Imperial College London, UK, hosted by the Computational Privacy Group at the Data Science Institute. He maintains an ongoing collaboration with deepkapha AI Lab on the same topics. He also has expertise in embedded systems design, including applications for AI, approximate computing and parallelization on multi-CPU/GPU systems.

Giulio Maspero physicist and theologian, is a professor at the Pontifical University of the Holy Cross. His research interests include the history of dogma, patristics (especially Gregory of Nyssa), and the relationship between philosophy and theology. He is a member of the steering committee of the Pontifical Academy of Theology.

Jie Mei received her M.S. in cognitive neuroscience from École Normale Supérieure Paris in 2015, and completed her Ph.D. in medical neuroscience at Charité—Universitätsmedizin in Berlin in 2019. Her major research interests include neuro-inspired AI, computational neuroscience, and applications of machine learning in neuroscience and neurology. She is currently based in Canada, where she heads AI research for deepkapha AI Lab and its subsidiaries. Her research interests include computational neuroscience, neurorobotics, machine learning and data analytics in healthcare and medicine.

Ionela Neacsu was an Assistant Professor at the Rennes School of Business and a postdoctoral researcher in the Strategic Management Department at the IESE Business School. She holds a Ph.D. in Business

Administration and Quantitative Methods from Universidad Carlos III de Madrid. Her research interests cover corporate governance, family businesses, strategic decision-making and executive compensation. Ionela has published in leading academic journals, including the *Journal of Management* and the *Human Resource Management Review*, and has presented her work at international conferences such as the annual meetings of the Academy of Management and the Strategic Management Society.

Dat Quoc Ngo is a senior computer science student at the University of Texas, Dallas, United States. As a research assistant he has spent the last two years researching machine learning at the intersection of neuroscience and speech. In his current full-time role at Surfboard, a social media start-up, he delivers big-data and machine-learning solutions to enhance data collection and advertising platforms for monetization. He is also a teaching assistant at LiveAI where he prepares study materials and provides study support to students. Dat is a visiting AI researcher at deepkapha AI where he researches natural language processing for medical issues. Dat's research interests are in unsupervised and reinforcement learning to solve the shortage of labelled data and language models.

Martina Nobili received her M.Sc. in biomedical engineering in 2020 and is currently pursuing a Ph.D. degree in Science and Engineering for Humans and the Environment at University Campus Bio-Medico of Rome, Italy. Her main research interests include open-source intelligence and multi-criteria decision aiding.

Gabriele Oliva received an M.Sc. and Ph.D. in computer science and automation engineering from the University Roma Tre, Rome, in 2008 and 2012, respectively. He is currently a tenure-tracked Assistant Professor in Automatic Control at the University Campus Bio-Medico, Rome. Since 2019, he has been an associate editor for the Conference Editorial Board for the IEEE CSS, and since 2020, an academic editor for the journal *PLOS ONE*. His main research interests include decision-making, distributed systems and optimization.

Francesco Pacileo is a research assistant at Sapienza University of Rome. He holds a Ph.D. in Commercial and Economic Law and lectures in Banking Law, Corporate Law and Extraordinary Transactions, and Listed Companies Law. He has been a speaker at several national and international conventions, and has published in prestigious legal journals as well as the book *Business Crisis: Going Concern and Solvency*.

Dimitris Paraschakis holds a Ph.D. in computer science from Malmö University, Sweden. His research interests centre around algorithmic and ethical aspects of AI, with a focus on recommender systems. Dimitris has published his research in recognized conferences such as RecSys, PAKDD, and IUI, and received a best-paper award for his early work on ethics-aware recommender systems. He acts as a programme committee member in several renowned AI venues (ECIR, RecSys). Dimitris remains active both in academic research as a visiting researcher at deepkapha AI Lab and in industrial AI practice as a data science consultant.

Cristian Righettini is a Ph.D. student in Oncological Science, Imaging and Technological Innovation at the Catholic University of the Sacred Heart, Rome.

Carlos Rodríguez-Lluesma is a Professor in the Department of Managing People in Organizations. He earned his first doctorate in philosophy from the University of Navarra and an M.B.A. from the IESE Business School. He also holds a Ph.D. in Organizations from Stanford University. Prof. Rodríguez-Lluesma has held various teaching and research assistantships at Stanford University, and has held lecturer positions at IESE Business School and the University of Navarra. He has also gathered valuable hands-on experience as a freelance consultant in such industries as financial services, biotechnology, consumer goods and management consulting. He serves as an adviser to a political consulting start-up in Silicon Valley.

Elvira Scarlat is an Assistant Professor in the Department of Accounting and Management Control at IE Business School. She holds a Ph.D. in Business and Quantitative Methods from Universidad Carlos III (Madrid, Spain). During her doctoral studies, she was a visiting researcher in the Accounting Department at Kenan-Flagler Business School, University of North Carolina at Chapel Hill. Her research interests are in the capital markets arena, with a focus on the interplay between corporate governance mechanisms and firms' financial reporting practices.

Malte Schmidt holds an M.Sc. degree in interdisciplinary business studies from Hanze University, Groningen. In his master's thesis he specialized in artificial intelligence in the renewable energy domain and has continued to work in this field. He is a senior business development lead and consultant for large IT and software enterprises in various industry verticals (financial services, utilities, insurance), with a keen interest in digital transformation, interdisciplinary thinking and applied business research.

Aileen Schultz is an award-winning legal innovator with a comprehensive career supporting the transformation of the legal industry. Her current role is in Thomson Reuters AI Innovation Lab. She is keenly focused on encouraging the adoption of advanced technologies within the legal sector, as well as the development of sound governance models for their ethical use across all sectors. She founded the World Legal Summit in 2018, bringing together perspectives from academic, government and legal institutions from over 20 countries.

Roberto Setola received his bachelor's degree in electronic engineering (1992) and Ph.D. in control engineering (1996) from the University of Naples Federico II. He is full Professor at the University Campus Bio-Medico, where he directs the Automation Research Unit and the Master's programme in homeland security. He was responsible for the Italian government's Working Group on Critical Information Infrastructure Protection (CIIP) and a member of the G8 Senior Experts' Group for CIIP. He has been the coordinator of several EU projects, and has authored nine books and more than 250 scientific papers. His main research interests are the simulation, modelling and control of complex systems, and critical infrastructure protection.

Tarry Singh is the CEO at deepkapha.ai. A seasoned tech entrepreneur, philanthropist and AI researcher, he has set up multiple start-ups. He has over 25 years' experience helping enterprises build and scale disruptive technologies to gain competitive edge. Through his brainchild deepkapha.ai, he aims to satisfy the world's hunger for AI with meaningful applications in business and society.

Roger Strand, Ph.D. is Professor of the Philosophy of Science at the Centre for the Study of the Sciences and the Humanities, and Principal Investigator for Team 4 Society, Centre for Cancer Biomarkers, both at the University of Bergen, Norway. He is Co-director of the European Centre for Governance in Complexity. Originally trained in biochemistry, Strand has a research focus on uncertainty and complexity at the interface between science, technology and society.

Ilaria Vigorelli holds a Ph.D. in philosophy, and her theological studies investigate the relationship between philosophical and theological discourse. She teaches Trinitarian theology at the Pontifical University of the Holy Cross in Rome, where she is also conducting a research project on relational ontology. An experienced trainer and coach, with a background in the analysis of mass communications, she has been associated with the FMV (Marco Vigorelli Foundation) since its birth, where she coordinates the commitment of her five brothers and fellow supporters.

LIST OF FIGURES

Chapter 6

LIST OF TABLES

Chapter 12

CHAPTER 1

Introduction

Marta Bertolaso⒟, Luca Capone⒟, and Carlos Rodríguez-Lluesma⒟

Abstract The chapter provides a general overview of the work undertaken by the contributors in the context of a series of meetings held in 2021. The method used in the writing of the volume and the criteria behind the structuring of the chapters are outlined. The working hypothesis behind the general concept of digital humanism is provided and its impact on the studied phenomena is described. A summary of the contents of the individual chapters concludes this Introduction.

M. Bertolaso (✉)
Campus Bio-Medico University of Rome, Rome, Italy
e-mail: m.bertolaso@unicampus.it

L. Capone
Campus Bio-Medico University of Rome, Rome, Italy
e-mail: l.capone@unicampus.it

C. Rodríguez-Lluesma
IESE Business School—University of Navarra, Madrid, Spain
e-mail: clluesma@iese.edu

M. Bertolaso et al. (eds.), *Digital Humanism*,
https://doi.org/10.1007/978-3-030-97054-3_1

1

Keywords Digital humanism · Ecosystem · Philosophy of technology · Artificial intelligence

This volume has been developed on the basis of a series of meetings between scholars and experts who share an ecosystemic and relational approach to the study of new technologies. Over the course of these meetings, a common methodological approach was established, which led to the contributions collected in this volume.

The contributions are divided into four sections, differentiated both from a thematic point of view and from the level of complexity at which the technological phenomenon has been analyzed.

This brief introduction will explain the working thesis and the methodological framework adopted.

1   Working Thesis

To develop an awareness of the breadth and articulation of the problems arising from new technologies, a reductionist approach is insufficient, no matter how multidisciplinary it may be. This means that the current utilitarian and hypercompetitive paradigm, although transversal and well beyond the economic sphere, remains reductive and unable to address the problems posed by the contemporary world, which asks for a systemic and relational viewpoint. This understanding emerged during the meetings as issues were addressed, such as, first, the coronavirus pandemic, but no less important, immigration, centralization of technological resources, human–machine relations and the replication of human cognition.

The new common framework to tackle these problems cannot come from a single discipline taking precedence over other fields, but must be the result of a real choral work. Complexity does not just mean dense articulation, nor does it imply a merely quantitative order of magnitude.

When we talk about complexity, we are referring to phenomena that cannot be reduced to a strict set of rules, that escape uniform reduction, and that exhibit asymmetries and nonlinearities, in ontological and epistemic terms. It is this definition of complexity that has caused us to abandon disciplinary divisions within the project. Problems related to several technological issues, such as the current pandemic, immigration and resource scarcity, require effort from all disciplinary fields. They cannot be solved by sensational discoveries or revolutionary inventions, but must be managed with time, patience, trial and error, admitting a pluralism (without relativism) of approaches that contribute to a unified understanding of the phenomenon itself.

For these reasons, this work should not be understood as a volume "on the digital". If this last element is separated from the form of life to which it belongs, the possibility of understanding its ecosystemic implications is lost, as well as the possibility of contributing to a common good.

The concept of ecosystem is understood along the same lines: it does not simply refer to the environment, but more deeply to human interactions from a systemic point of view. In this sense, the environment, and thus ecosystem dynamics, are understood in terms of green, blue (digital) and white (organizational) interactions.

Thus, before talking about technology, this book is about the human being. It is to this peculiar subject that all sections of the book are dedicated and only through this lens will the digital phenomenon be investigated. The human subject is a complex and historical being, and like contemporary problems, it requires to be studied under different levels of abstraction. In the volume we propose four layers of increasing granularity.

The main theses of the volume are as follows:

- Technologies constitute the inescapable ecosystem within which human action takes place. They do not constitute a mere material substrate but provide the background of meaning within which individual and social action is constituted as such. In this sense, they are essentially non-neutral.
- The human being is a relational individual in a double meaning: he is in relation with others but he is also in relation with his environment, both natural and artificial.
- The affirmation of this relationality precludes a determinist approach to human action and opens up a scenario of uncertainty and complexity within the technological environment.
- This same complexity makes human actions free and autonomous, despite technological mediation. In the same way, the technological processes involved in this mediation should not be framed in terms of irreversibility but of dialectics between subjects and environment.

Only from this perspective do the problems related to issues of complexity, uncertainty, the social and individual impact of technologies, ecosystem sustainability of technological progress, and the autonomy of the enhanced subject become understandable.

Through the levels of abstraction outlined, and pursuing a human-centric approach, the book offers a network of viewpoints and an up-to-date reflection on foundational issues about both emerging understandings of the digital era and still hidden and ignored aspects that could instead be dramatically relevant in the future, in the process of a digital humanism.

2   Methodological Framework: Digital Humanism

The term digital (or more generally, technological) humanism represents a hybrid approach to digital technologies at the intersection of the relationships between subjects and the technological environment. This concept should be seen, not as an idea to be pursued until a digital humanism has been established, but as a method that is differentiated from modern reductionism and avoids the aporetic conclusions of post-modernism. Within this framework, digital technology is here studied through the relationships that it entertains with the human being.

The presentations and debates that took place in the three meetings around the concept of digital humanism made it possible to identify four main directions along which the discourse on new technologies must develop in order to formulate a multidisciplinary theoretical framework.

Each of these viewpoints constitutes a precise level of abstraction for framing the digital phenomenon. At the same time, they are not isolated points of observation but stacked layers that influence each other. The levels (Fig. 1) should be interpreted as levels of abstraction with increasing granularity. The layers range from a maximum level of generality to an increasing focus on the particular phenomenon.

3   Ethical Issues

Although the above considerations relate to the ethical aspects of digital technology, the main aim of the book is to provide an overview of the basic questions, theoretical issues and hot topics raised by technology. If McLuhan wondered about the role and destiny of writing in the age of the TV, of the telephone and of the large-scale mobilization of people,

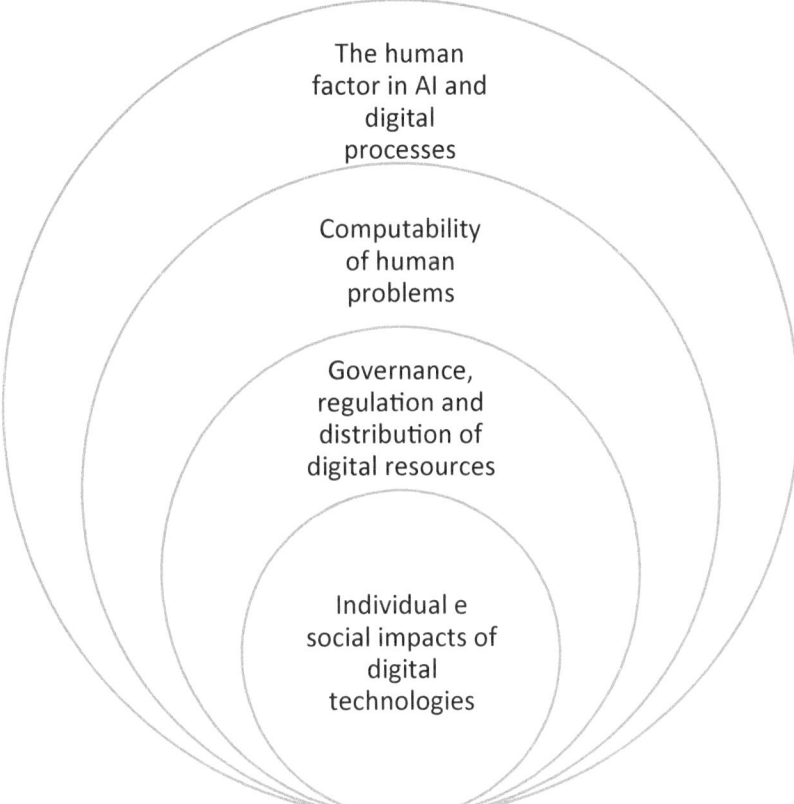

The human
factor in AI and
digital
processes

Computability
of human
problems

Governance,
regulation and
distribution of
digital resources

Individual e
social impacts of
digital
technologies

Fig. 1 Levels of abstraction

today we are also faced with a major rethinking of the institutions with which we ordinarily have to deal. Just think of how the pandemic has accelerated technological changes in education and in the workplace. It will be necessary to understand how other legal, economic, business, entrepreneurial and scientific institutions will modulate their aims and methods of action in the light of the changes taking place. So, if this work is about ethics, it is about a common ethics, not individual but social. Something that looks less at the individual action and more at the network of relations within which it takes place.

Thus, on the one hand, the book can be imagined as a series of perspectives on scenarios and strategies set in motion by recent technologies; on the other hand, the approach that holds together these perspectives belonging to the same heterogeneous and intricate subject matter remains fundamental. The cardinal assumption of this work remains the systemic and relational epistemological paradigm necessary to delineate the most relevant problems posed by digital technological progress and AI.

4   VOLUME STRUCTURE

Part I of this volume investigates the general problems posed by artificial intelligence and digital technologies. Consequently, this section includes the most philosophical and theoretical contributions.

At this level of analysis, practical problems give way to epistemological issues, to the control of definitions and to the drafting and outlining of a general theoretical model that will be the background to the contributions in the other parts of the volume. This model, based on a relational paradigm, must rely on the linguistic aspect of the concept of relationship—of the human being with his fellows, but also with the world and experience. The linguistic relationship brings to the forefront gnoseological issues related to the replication of human cognition, or rather intelligent behavior.

In addition, the first part of the book deals with general problems of epistemology of technology, scientific reflection and common sense on the digital and its contemporary implications.

All the topics addressed are related to language in various ways. In this sense, one of the main human factors within technologies is precisely language, that peculiar element that has fascinated scholars since ancient times and has often been regarded as a distinctive element of sentient beings, compared to animals without speech.

Not only does language technology distinguish humans from the rest of the living world, it also makes progress possible. Language enables the storage and transmission of knowledge, as well as its systematization, giving rise to science and technology. Language is also the fundamental element that enabled human beings to live in communities and to organize social life on an ever-increasing scale.

While language is one of the pivotal tools for human success, it contains the seeds of its degenerate and self-referential use. The modern outcomes of the epistemology and reflection on technologies illustrated by Latour are widely exposed in the chapter 2, linking their premises to the digital question.

Capone and Bertolaso lay the epistemological foundations for a philosophically informed understanding of artificial intelligence and in particular of one of its branches, currently the most important—deep neural networks. The authors, relying on Latour's main work, *We Have Never Been Modern*, address the rationalist reductionism towards technologies. Following Latour's reasoning, the authors apply the anti-modern theoretical framework to deep neural networks. This critical section is followed by a discussion in which the importance of human representational methods for implementing and training neural networks is illustrated. This refutes reductionist and anthropomorphic biases related to artificial intelligence.

In Chapter 3, Vigorelli, Malo and Maspero take the issue of the methods of representation to the next level. The authors focus on the linguistic side of the relationship between human beings and the world. If the gnoseological relationship is inaugurated by language, we shall see that it is not exhausted by it and in what way the living being differs from the machine in its relations.

In Chapter 4, Strand and Kovacic return to Latour's argument. The chapter draws the most practical and perhaps most dramatic conclusions from the reductionist, or modern in Latour's words, attitude towards technologies and their use.

Part II, in contrast with the first, theoretical, set of chapters, addresses the issue from a practical point of view. In its three chapters, contemporary problems are illustrated, and AI-based practical solutions are proposed. While Part I warned against an entirely computational and rationalist conception of the technological approach to human problems, Part II seeks to illustrate how some problems are actually computable and solvable by means of technology.

The authors of Chapter 5, Nobili, Oliva and Setola, try to combine machine learning techniques with decision-making processes in real-world scenarios, combining technology with the experience of human actors, decision makers and subject matter experts.

In Chapter 6, Ali, Ngo, Bhattacharjee, Maiti, Singh and Mei address the issue of depression detection. In the context of the increased incidence of mental health disorders due to pandemic lockdowns, the authors seek to develop tools for NLP-based depression detection and show how a pre-trained model can be fine-tuned to detect depression.

Chapter 7 looks at the application of artificial intelligence to the energy transition, with a particular focus on photovoltaic systems. Schmidt, Marrone, Paraschakis and Singh, focusing on the Dutch context, report on the maturity level of the solar PV industry and suggest strategies for improving the situation.

In this part, as in the first, technology is studied only through the lens of human beings and their relationship with the environment, a paradigm defined in this volume's title as "digital humanism". Chapter 5 presents the relational component at the level of the concrete experience of decision makers, supported by technological tools; in Chapter 6, language returns to the forefront through the adoption of the transformer model for depression detection, and with it the pivotal tool of the human relationship; and in Chapter 7, sustainable AI solutions to environmental problems, decisive for the future of man's relationship with his ecosystem, are explored.

The title of Part III, Governance, Regulation and Distribution of Digital Resources, is based on a concept that runs through all its contributions: the concept of *ecosystem*. The governance and regulation of technologies must be integrated and ecosystemic, just as the distribution of digital resources must be human centric and distributed. The concept of ecosystem does not apply on a merely superficial level but incorporates theoretical and practical issues. An ecosystemic conception of technology-related phenomena is a prerequisite for their proper regulation.

Chapter 8 discusses the legal recognition of technologies by reintroducing the question of dualism already found in Latour, and aims to provide a different understanding of technologies. The chapter illustrates different conceptions of natural and technological phenomena based on their legal recognition. Migle Laukyte clearly illustrates how these conceptual problems have serious repercussions for policy and regulation, and for the coexistence of the technosphere and the biosphere.

Consistent with the ecosystemic approach, Chapter 9 tests the possibility of reframing the concept of legal identity, detaching it from the geographical and national context, making it an inalienable right of every human being. The distributed nature of technologies provides the means to do so. Schultz and Di Giulio explore the global identity challenge by proposing a distributed governance framework, starting from specific cases such as that of Estonia.

The last contribution of this part, by Francesco Pacileo, addresses the discrepancies within European regulation on high-risk and non-high-risk AI. The article takes the form of a case study highlighting the consequences of neglecting the holistic approach to technology.

The fourth and final part, Part IV, focuses on the individual and social impacts of digital technologies. The three contributions, from different points of view, seek to understand the way in which technologies shape, for better or for worse, different areas of social life, looking at them from an ecosystemic perspective. The different areas observed are conceived as micro-ecosystems in a systemic relationship with each other.

The chapter that opens Part IV shows the need for a pedagogy that is up to date with technological progress, and is capable of educating people in the use, but also in the understanding, of technological tools. The pedagogical project envisaged by Malavasi and Righettini is not limited to the field of early education, but extends to adult education, with a constant tension towards digital humanism.

Chapter 12 sheds light on how ICT shapes the corporate world and in particular TMT (top management team) agility. If the ecosystem and relational approach is the most suitable way to frame digital phenomena, the workplace is also a complex environment, which can be shaped by technology along different dimensions.

Chapter 13, Digital is the New Dimension, gets to the heart of the matter, bringing together the particular and the general. If technology is ecosystemic, if it is like an environment with the human being at its center, then digital is the new dimension. This dimension is clearly shaped by the technologies that structure it and by those who have control over these technologies. In this sense, the authors, Fung and Lim, stand in the middle between an entirely instrumental conception of technologies and a vision that sees technology as an autonomous force capable of directing human action altogether. The chapter is a dense pathway comparing cultures, arts and traditions in the context of the new dimension that has been outlined in the course of the discussion.

The Human Factor in AI and Digital Processes

CHAPTER 2

Reflections on a Theory of Artificial Intelligence

Luca Capone and *Marta Bertolaso*

Abstract In the following pages, Bruno Latour's conception of modern epistemology is analyzed. Although Latour considered modernity to be at an end, the chapter shows how the modern analytical approach is still active in the contemporary world.

Research on artificial intelligence (AI) is no exception. Despite the fact that some experts and practitioners continue to promulgate a modern and reductionist conception of technology, the chapter show how that

---

This contribution was conceived in a unitary way The first, second and sixth paragraphs were drafted by Marta Bertolaso, and the third, fourth and fifth by Luca Capone.

L. Capone (✉)
Campus Bio-Medico University of Rome, Rome, Italy
e-mail: l.capone@unicampus.it

M. Bertolaso
Campus Bio-Medico University of Rome, Rome, Italy
e-mail: m.bertolaso@unicampus.it

M. Bertolaso et al. (eds.), *Digital Humanism*,
https://doi.org/10.1007/978-3-030-97054-3_2

approach does not stand the test of time and how it leads to misunder-standings and misconceptions about AI and human cognition.

By reflecting on the theoretical assumptions underlying neural networks, currently the state of the art in AI research, the paper proposes an alternative approach to epistemological reductionism. These assumptions bring Latour's call about the constructed and hybrid character of phenomena back to the centre of the debate. This approach will prove fundamental in laying the groundwork for a philosophical reflection on AI.

Keywords Philosophy of science · Epistemology · Philosophy of language · Philosophy of technology

1   INTRODUCTION

In 1991, Bruno Latour published his famous work, *We Have Never Been Modern*. Confronted with a world made up of phenomena that he described as hybrids (neither natural nor cultural), such as the ozone hole, the AIDS tragedy, world economic and political crises, and the first decisive steps of computers among the mass media, the author proclaimed himself a witness of the end of modernity, or rather of the illusion of modernity. Today, 30 years later, the hybridization of the natural and the cultural, the technical and the biological, the subjective and the objective, continues, but it is not entirely certain that the modern illusion has ceased (Latour, 1993, pp. 49–50).

Nowadays, artificial intelligence (AI) has become the perfect representation of contemporary hybrids, yet many of its implementations and interpretations are undermined by a purely modern conception of cognition and technology. In the following pages this epistemological background is illustrated and some possible ways to avoid its shortcomings are suggested.

The second section deals with modern epistemology in general, as formulated by Bruno Latour, and the third illustrates the particular application of modern epistemology to the question of AI systems. The fourth section provides a different interpretation of these systems, relevant to their understanding and development. In the fifth section, the question of the relationship between artificial intelligence and human cognition is addressed. Finally, the sixth section provides some remarks on the future developments of AI in the light of Latour's reflection.

2   Modern Epistemology

Latour recognizes the deeply hybrid character of the world of his time, a world made up of objects with heterogeneous contours, not entirely natural, but neither entirely cultural, not mere objects, but not real subjects either. The hybrid character of the phenomena reported by the author is expressed in dichotomies such as those just mentioned. The modern human being is placed in this context (Latour, 1993, pp. 51–52).

The sociologist defines modernity primarily as an intrinsic need for analysis. The modern observer is inevitably bound to a process of selection of phenomena. This analytical procedure extracts the object of analysis from the network of relations in which it is found, and places it under the particular observational lens of a discipline. In this way, the object is abstracted from its hybrid nature and secured on one side of the dichotomy, nature or culture. This process of abstraction raises important epistemological questions, since it imposes a special constitution on phenomena that are instead subject to the laws of many different domains. For Latour this is unacceptable—the modern processing of hybrids is inadequate to describe contemporary phenomena (Latour, 1993, pp. 2–3).

A contemporary example of a hybrid is represented by AI systems. That of AI is a scientific problem, which includes various disciplines: mathematics, statistics, computer science. In addition, any particular application of AI involves other disciplines, such as medicine, finance and law. Eventually, AI research will involve philosophical questions about language, cognition and sensibility in the fields of natural language processing (NLP) and computer vision. Lastly, there are technical and logistical questions about data storage and hardware resource, as well as national security concerns, competition among states, and environmental and sustainability issues. The phenomenon of AI spreads across the network of relationships constituted by contemporary society, from African mines for the extraction of minerals needed for computers (Amnesty International, 2016), through cables on the bottom of the ocean, to university laboratories and newspapers reporting on their latest findings. Latour sees this kind of complexity in most of the phenomena he observes.

In opposition to the critics (those who carry out the analysis of phenomena), Latour identifies a class of scholars that includes economists, historians, philosophers and sociologists; in the author's view, the task of these disciplines should be "retying the Gordian knot" cut by modern thinking, reconnecting nature and culture. However, Latour finds the symptoms of modernity in these fields as well (Latour, 1993, p. 3). Epistemology finds itself caught between a naturalizing reductionism and a relativistic constructivism. The former believes that it can treat phenomena in isolation, the latter that it can cut off the referent, focusing on the discourse's truth effects.

At the end of his diagnosis, Latour argues for an alternative to modern epistemology that can mend the gap between natural and cultural, subjective and objective, bare fact and representation.

His project does, however, encounter two obstacles:

1. In its work of analysis, modern culture acts by removing the hybrid origin of phenomena. This is the main reason why modern realism fails to look beyond its objects, back to the web from which they were extracted.
2. A critique of modern analytical methods invariably suffers from an accusation of hypostatizing the representation of phenomena, the discourse, while ignoring referent and speaker, leading to a sterile relativism.

Because of these two tendencies, modern intellectual life takes place within the polarity between naturalized facts and truth effects. The next section illustrates how this epistemological dichotomy manifests itself today in the field of AI.

3   AI as a Hybrid

To summarize, the word *modern* defines two sets of practices: the first set, called *translation*, creates a mixture; the second set, called *purification,* produces ontological areas, human and non-human, natural and cultural, subjective and objective. The first set corresponds to networks, the second to criticism. The former links in an endless chain: atmosphere, strategies of science, industry, states, ecology; the latter establishes a separation between nature and culture, simultaneously removing the traces of the operation (Latour, 1993, p. 10).

The result is, on the one hand, the concealment of the social origin of natural objects, a removal of the representative mechanisms, the technical apparatuses, the institutional, political, cultural and historical subjects that constitute the objects that undergo the modern naturalization; on the other, the development of a relativistic and integrally conventional conception of cultural phenomena.

While Latour foresaw the end of this modern illusion, it must be noted that the process of disenchantment is far from complete. According to the sociologist, the proliferation of hybrids he was witnessing was the condition for the illusory nature of modern epistemology to be revealed. Nowadays, however, the same technological hybrids have been incorporated into the modern mentality, going through the process of naturalization described above and almost becoming an object of veneration. Technology, like any cult, has its ministers.

An example of this attitude is expressed by Chris Anderson in his famous article on the future of science in the age of AI (Anderson, 2008).

The End of Theory?

Chris Anderson is a British-American author and entrepreneur, former editor-in-chief of *Wired* and currently CEO of 3D Robotics. In a famous article in 2008, he proclaimed the end of theory and the obsolescence of the scientific method. Their place would be taken by AI systems.

Anderson proposes several examples where statistical methods, coupled with large amounts of data, were able to solve problems that were challenging for researchers. Based on such cases, he foresaw the gradual extinction of traditional scientific methodologies in favour of data-driven research based on the construction of probabilistic models and big data.

In Anderson's modern conception of science and technology, scientists formulate models that "are systems visualized in the minds of scientists" (Anderson, 2008). Once formulated, the models are then tested with experiments that verify or falsify them, deciding the fate of the theories.

This is undoubtedly a very romantic view of scientific work, but not a very objective one. Science has always been a matter of theories, of course, but also of tradition, of the history of scientific thought and its methods, of the institutions that promote it and of the available resources, of the good fortune of researchers, and finally of technology. Anderson does not compare a data-driven research model with science, but with a trimmed-down version of it.

However, the most important aspect is that the conception of technology expressed in the article is similarly biased. Technology has always been part of science and, before that, of human experience in its ordinary sense (think of the first numerical notations and the invention of writing; see Schmandt-Besserat, 1996).

Moreover, Anderson seems to believe that AI functions as a kind of unerring eye, which is able to see patterns in data that are not accessible to humans, something like the intimate rationality of phenomena. As he says, just provide "the data" and the phenomena are explained, without the need for any theory. All kinds of phenomena can be modelled, from natural to social ones. New animal species can be discovered with gene sequencing, stock market trends can be predicted by analyzing the social activity of certain user groups, medical diagnoses can be provided, and even Mandarin can be translated into Spanish and then back into English. All this without knowing biology, sociology, economics or languages.

Although appealing, this description of how AI works is misleading, and it shifts the focus from the phase of data representation and structuring (the hybrid origin of phenomena), to that of their modelling. In this respect, Anderson shows that he fully adheres to the modern spirit criticized by Latour. On the one hand, he embraces a metaphysical realism whereby representation, or theory, does not play an active role in the understanding of phenomena. Phenomena are conceived as given a priori, and models faithfully reflect them. Symmetrically, theories, as part of culture, are relative, methods of representation, expedients that can be dispensed with.

Latour would have been disappointed, had he read these statements, and would have certainly considered it appropriate to investigate whether it is actually possible to proceed with scientific research in the manner proposed by Anderson. To do so, it is necessary to understand what kind of object neural networks are. This is the argument covered in the next section.

4   Deep Neural Networks

Indeed, neural networks have something common with scientific research work. Part of scientific and theoretical work deals with the formulation of mathematical models that describe the investigated phenomena. A neural network does something very similar—it can be defined as a parametric function (a mathematical model) in the form of a computational graph (Fig. 1).

The network consists of an input layer, an output layer and internal layers. The inner layers are divided into weights and neurons. A neuron is essentially a small parametric function (linear or non-linear), where the neurons constitute the invariable part of the function (network). The weights, on the other hand, are the parameters of the function, its variable part. "Each weight is like a volume control knob that determines how much the next node [neuron] in the graph [network] hears from that particular predecessor in the graph [its input]" (Russell & Norvig, 2021, §21.1.1).

During training, the network is provided with inputs (data) and outputs, through which it calculates the parameters of the function. These are first randomly initialized, and the network tries to predict the output for each specimen in the train set. At each cycle the network compares the prediction with the given correct output, it calculates the error and updates its weights accordingly. After training, during tasks execution, the network uses the trained parameters to calculate the outputs based on inputs never seen before.

In spite of Anderson's statements, it is not true that there is no theory behind deep neural networks (DNNs). The basic assumption of these

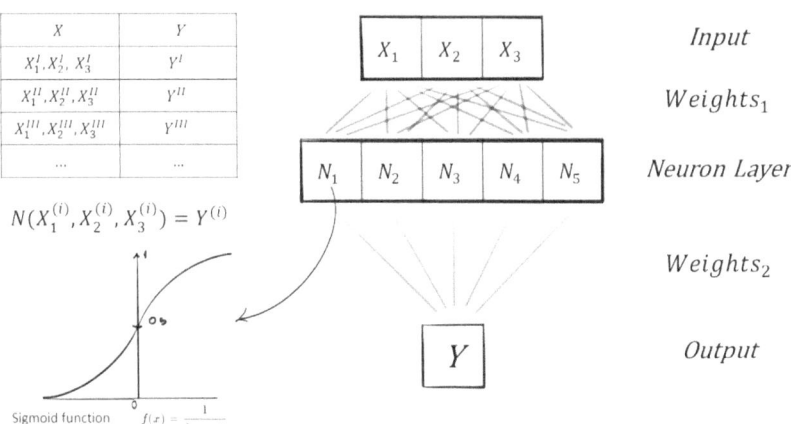

Fig. 1 A simple neural network

systems is the universal approximation theorem, which states that a sufficiently large function with at least one non-linear and one linear layer is able to approximate any continuous function with a certain degree of accuracy (Russell & Norvig, 2021, §21.1.1). In a nutshell, it is able to find a mathematical model for any data distribution.

There is another common aspect shared by neural networks and scientific research that is ignored by Anderson's description, namely the representation of phenomena. Data are not neutral entities, given a priori. The data are the product of a representative process to which the object is subjected, whether it is a written description or a measurement. Moreover, there are no isolated representations (isolated data), but always systems of representations. A piece of data only makes sense if it can be compared with other similar pieces of data, if it has relationships with other represented objects and if it is embedded in a general practice. In short, the data form a differential system with each other, in which the value of each piece of data is relational, based on the relationships with other elements in the system.

This applies regardless of whether one is dealing with sophisticated phenomena or ordinary objects. The simplest example of this is the economic representation of the value of an object. It is possible to say that an object costs x, but in order for this piece of data to be informative in some way (for example, to know whether x is a lot or a little) it must be possible to compare the data with analogous representations of other objects. The same applies to more sophisticated representations such as linguistic ones. The meaning (value) of a sign depends on the system of signs in which it is placed (see de Saussure, 2011, p. 115).

To represent an object means to place it within a system of relations based on certain relevant traits such as monetary value, physical extent (think of the prototype meter in Paris,[1] see Wittgenstein, 1968, §50) or temperature (based on the freezing and boiling points of water at 1 atm pressure), while neglecting others. The process of traits selection and

---

[1] The meter prototype, housed in Paris, is a platinum bar used as a standard for the international metric system from 1889 to 1960. The current standard is the one promulgated by the Geneva Conference on Weights and Measures, which defined the meter as the fixed numerical value of the speed of light in vacuum c to be 299,792,458 when expressed in the unit m s^{-1}, where the second is defined in terms of the caesium frequency Δv_{Cs}. https://www.bipm.org/en/si-base-units/metre.

representation are essential elements of gnoseology and a relevant part of the theorizing process in science.

Neural networks are no exception. In order to formulate probabilistic models and make predictions, they must model relationships between entities represented according to predetermined criteria, based on a certain worldview. Far from dispensing with theory, neural networks need human representational systems to be able to *see* their objects (Capone, 2021).

An example may be useful. The neural networks have two techniques for task solving: regression and classification. Regression (Fig. 2), starting from certain features, interpolates the missing data, for example, a network can be trained to estimate the price of a house based on certain features (advertisement, size, year, area, etc.).

A classification system (Fig. 3), on the other hand, groups its objects according to whether they belong to a certain category. An algorithm can classify images according to the object represented. In both regression and classification, in order to perform the task at hand, the network places its objects (houses and images) in a system of relations.

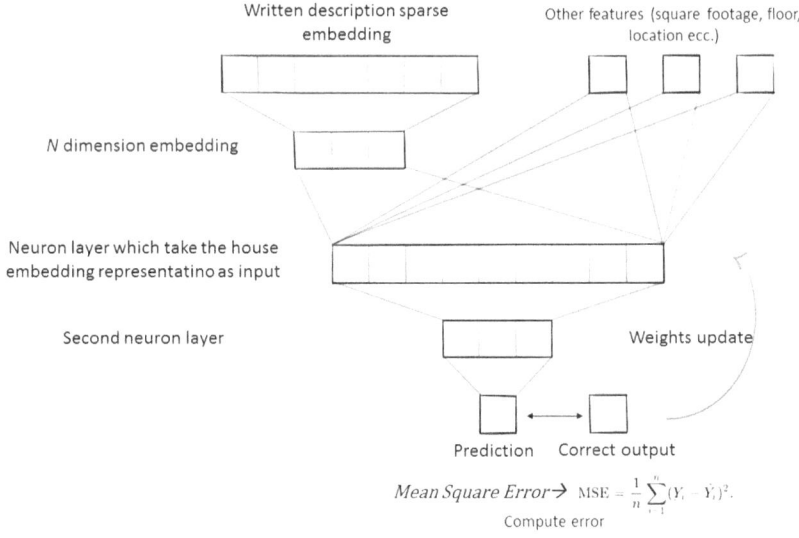

$$\text{Mean Square Error} \rightarrow \text{MSE} = \frac{1}{n} \sum_{i=1}^{n} (Y_i - \hat{Y}_i)^2.$$

Fig. 2 A neural network for regression tasks

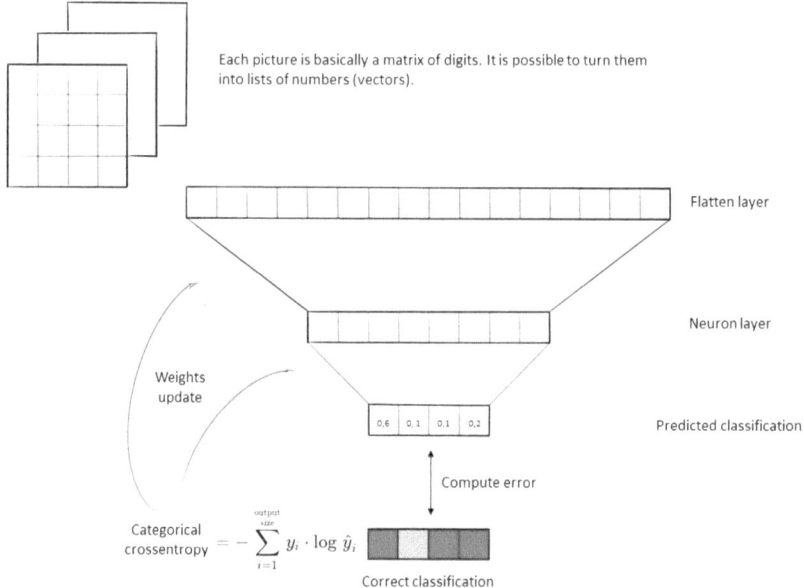

Fig. 3 A neural network for classification tasks

In both cases, the input layer passes the features of the first specimen to the first layer of parameters and neurons. These, based on the data, generate an embedding of the object with *n* values. An embedding is a numerical array (a vector) that represents the object.

It is possible to graphically imagine the embedding numbers as coordinate values in an *n*-dimensional space. Once calculated by the network during the training phase, these values correspond to a position in the network space in relation to the other elements of the train set. In general, vectors are composed of hundreds of elements, so they are not visualizable, but through techniques such as PCA (principal component analysis) it is possible to reduce their dimensionality for illustrative purposes (Fig. 4).

Clearly, it is not possible to specify what the single element of the embedding represents, since all the information provided to the network during training (features and correct outputs) is condensed into it. What is important is that the network provides a topological (see Rasetti, 2020), or relational, representation of its elements.

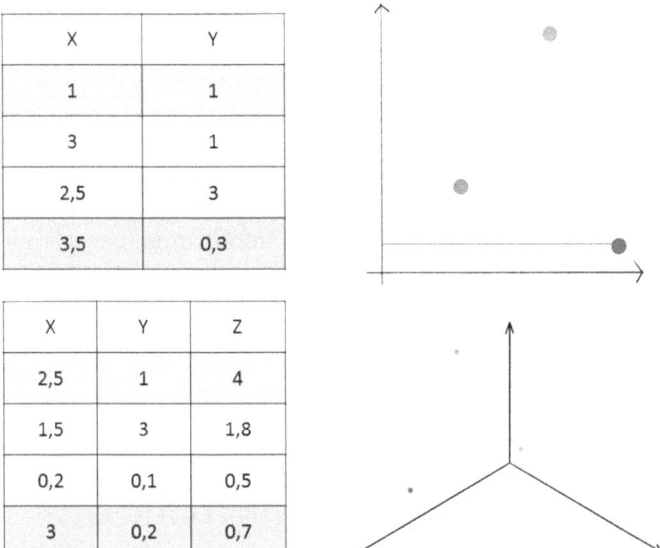

Fig. 4   2- and 3-dimensional embeddings projections

These elements, in addition to not possessing intrinsic properties, but only relational ones, embed in their representations a lot of collateral information present in what at first sight might seem to be neutral data. An algorithm that estimates house prices might be affected by a housing bubble and represent it in its embeddings; a system for assisted diagnosis, for curriculum analysis or for mortgage approval might have sexist or racist biases, acquired from train-sets (Bolukbasi et al., 2016). Although there are methods to mitigate these undesirable effects, the representational method of neural networks makes it quite clear that we are not dealing with neutral objects, but that the representations encompass the whole ecosystem of relations woven by the data, and that it could not be otherwise.

In conclusion, what seems most surprising to Anderson is the network's ability to provide answers without knowing what it is calculating. But this is a misleading statement. This chapter resolves this issue by proposing three preliminary conclusions:

1. What is most important in the training of neural networks is the representation of the data. The data must be represented in such a way that the network can calculate its parameters properly.
2. The representation must be established from the very beginning; it is the structure, or the set of categories, that the network must take into account. In this sense the representation can already be understood as a theory.
3. Representations incorporate a lot of information. On the one hand, they reduce the object by selecting relevant traits. On the other, these traits do not return neutral values, but the rich world of relations that objects hold with the others, from a certain perspective.

Now that the modern conception has been set aside, it is possible to consider the question of the status of AI more comprehensively.

5   REPRESENTING INTELLIGENCE

In the previous section, the hybrid character of AI was illustrated. In dealing with data, the modern mentality believes that AI systems can isolate a piece of pure world, from which it can derive the intimate rationality of phenomena, but this is not the case. It was shown how data are intimately linked to the concept of representation and how, because of this, they are both poorer and richer than the bare facts. Poorer because they select a slice of reality according to certain criteria; richer because they do not simply return features on the isolated object or phenomenon but place it in a system of relations that includes all the others. All that remains for the algorithm to do is to calculate a model of their distribution. Neural networks calculate this distribution providing vector representations of the analyzed elements, ideally placed in an n-dimensional space. The topology of this space, the reciprocal relationships between objects within it, returns a model of the studied phenomenon.

This raises a question that is not found in Latour's book, but has always been fundamental to AI studies, namely whether it is possible to define these algorithms as intelligent and to what extent.

The Question of Intelligence

From its very beginnings, AI research has pursued the replication of human thought as one of its aims. Alan Turing (1950) was among the first to speculate about this issue, in his famous essay entitled "Computing machinery and intelligence", formulating the *argument from informality of behaviour*.

According to Turing, human behaviour (understood, in a broad sense, as intelligent behaviour) is far too complex to be captured by a set of formal rules. Turing does not intend to deny that the issue of consciousness is a concrete and important problem, he simply does not consider it relevant for AI research (Russell & Norvig, 2021, §27.2.2).

Turing's approach has been supported by a number of people over the years. An effective argument in its favour put forward by Edsger Dijkstra (1984) states that asking whether an algorithm *thinks* or *knows* what it calculates is as relevant as asking whether a submarine swims. The question of algorithmic intelligence is ill-posed and is once again the result of the modern approach. The modern attitude tends to isolate thought, in order to make it easy to investigate, while losing the context in which this phenomenon takes place.

It must be accepted that intelligence is not a defined and homogeneous thing as one might imagine, but something both internal and external, discursive and material, biological and cultural, individual and collective, a practice that cannot do without speech and sign manipulation (phonic or graphic), social cooperation, community living, tradition, and a biological endowment that has little or nothing to do with the computational neuron.

Since it is not possible to define an essence of intelligence, the most correct question is the one posed by Turing, namely whether it is possible to reproduce intelligent behaviour.

Intelligence as a Complex Heterogeneous Phenomenon

It has been claimed that intelligence is not a bounded and computable phenomenon, and it is not by chance that Turing spoke of behaviour. Rather, intelligence can be described as something hybrid, branched, which is part of biology, culture and the materiality of the objects with which living beings interact. If it is not possible to place oneself on the general and heterogeneous plane of intelligence, it is nevertheless possible to model certain areas of it. In this regard, one of the most developed AI domains in the last ten years has been NLP.

This is not surprising, since cognition is characterized by an important linguistic component. It is possible to state that the operational, pre-verbal thinking found in children only comes to maturity thanks to the intervention of language and specifically the ability to organize one's own behaviour through material signs (phonic and later graphic) (Vygotsky, 1987, p. 114).

Since language is a system of signs, a mathematical model is able to capture its distribution, and consequently its semantic background. Such a language model provides the machine not only with a linguistic competence, but something similar to a map of language meanings, a model of the relevant uses of concepts, a sort of topology of meanings.

A pre-trained language model such as GPT-3 (Brown et al., 2020) or Bert (Devlin et al., 2019) is able to answer questions in a coherent manner, to pronounce on subjective questions and to make judgements. This is due to the hybrid character of cognition and its blending with language.

If these models were based on logical hierarchies of concepts and rules attempting to define the structure of thought in its entirety, none of this would be possible. On the other hand, word embeddings, by placing signs in a vector space (Fig. 5), make it possible to perform operations between signs (Mikolov et al., 2013) and to ask the machine questions in natural language. Clearly the machine does not have *thoughts like we do*, but words representation based on large amounts of written text

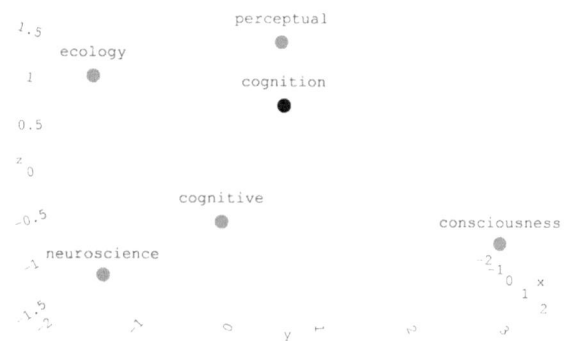

Fig. 5   Five nearest-neighbour embeddings of the token "cognition"

allows systems to formulate rich representations of concepts that incorporate judgements, similarity and dissimilarity of meanings, appropriateness according to semantic context and much more.

The 300 value vectors corresponding to the tokens have been reduced to three dimensions by PCA.

From a theoretical point of view, AI systems take strong positions on language and cognition. In the case just presented, following distributional semantics and structural linguistics, language is conceived as a differential system of signs in which the semantic value of each element is determined by the positionality of each sign within the linguistic system (see Harris, 1954; de Saussure, 2011). Moreover, based on this theory, some strands in the field of psychology relate thought to a predominantly discursive activity, in which concepts are also defined thanks to the differential (categorizing) activity of discourse.

This is not the place for a theoretical analysis of semiotic systems. What is important is to investigate the possibility of a modelling of hybrid phenomena, such as human cognition, thanks to digital technologies, as opposed to reductionist failures due to a modern understanding of phenomena (Dreyfus, 2007).

At this point, one has to ask whether NLP algorithms really succeed in reproducing the hybrid phenomenon that is thought. The most honest answer is "not yet". The model under consideration is undoubtedly able to reproduce only a small part of the complexity and richness of human thought—nevertheless, a precious part. After many years of research, a truly non-modern and hybrid approach to the reproduction of cognition is manifested for the first time, an approach that does not seek to extrapolate idealized objects, but which blends with the materiality of the human world. Language, far from being a mere code for communication, is a tool that structures society, whose meanings incorporate salient information about human practice. This has positive implications, since linguistic meanings enable NLP algorithms to work with concepts in an appropriate way, but it also has negative implications, since meanings incorporate biases and prejudices that affect the models that represent them (Capone & Bertolaso, 2020).

6   Conclusions

In the previous sections, Bruno Latour's conception of modern epistemology was explained. Although Latour considered modernity to be at an end, it was shown how the modern analytical approach is still active in the contemporary world. Research on artificial intelligence is no exception.

Despite the fact that some experts continue to promulgate this interpretation,[2] the results of the research show how that approach does not stand the test of time and that the modern approach leads to misunderstandings and misconceptions about AI and the replication of human cognition.

The point is not that the algorithms can give answers without knowing the meaning of the questions, but that the representations of the objects provided by humans in the training phase already contain more information in themselves than an analytical approach would capture in them.

Although the full replication of human cognition is still a long way off, the recognition of the hybrid and complex nature of these phenomena is a step forward. In this sense, a philosophical approach to the problems of AI proves to be of fundamental importance.

REFERENCES

Amnesty International. (2016). *Is my phone powered by child labour? Behind the flash screens of our smartphones lurks a secret that phone manufacturers would rather ignore.* Retrieved September 19, 2021, from https://www.amnesty.org/en/latest/campaigns/2016/06/drc-cobalt-child-labour/

Anderson, C. (2008). *The end of theory: The data deluge makes the scientific method obsolete.* Retrieved September 19, 2021, from https://www.wired.com/2008/06/pb-theory/

Bolukbasi, T., Chang, K. W., Zou, J., Saligrama, V., & Kalai, A. (2016). *Man is to computer programmer as woman is to homemaker? Debiasing word embeddings.* https://arxiv.org/pdf/1607.06520.pdf

Brown, T. B., Mann, B., Ryder, N., Subbiah, M., Kaplan, J., Dhariwal, P., Neelakantan, A., Shyam, P., Sastry, G., Askell, A., Agarwal, S., Herbert-Voss, A., Krueger, G., Henighan, T., Child, R., Ramesh, A., Ziegler, D. M., Wu, J., Winter, C., … Amodei, D. (2020). Language models are few-shot learners. arXiv:2005.14165 [cs].

Capone, L. (2021). Which theory of language for deep neural networks? Speech and cognition in humans and machines. *Technology and Language, 3*(2). https://doi.org/10.48417/technolang.2021.03

Capone, L., & Bertolaso, M. (2020). A philosophical approach for a human-centered explainable AI. *Proceedings of the Italian Workshop on Explainable Artificial Intelligence Co-located with 19th International Conference of the*

---

[2] Noam Chomsky has recently been critical about AI and in particular NLP (2019).

Italian Association for Artificial Intelligence (AIxIA 2020), 2742(80–86). http://ceur-ws.org/Vol-2742/short1.pdf

Chomsky, N. (2019). *Deep learning is useful but it doesn't tell you anything about human language*. Interview by Lex Fridman. https://www.youtube.com/watch?v=ndwIZPBs8Y4&ab_channel=LexFridman

de Saussure, F. (2011). *Course in general linguistics* (P. Meisel & H. Saussy, Eds.; W. Baskin, Trans.). Columbia University Press.

Devlin, J., Chang, M.-W., Lee, K., & Toutanova, K. (2019). Bert: Pre-training of deep bidirectional transformers for language understanding. arXiv:1810. 04805 [cs].

Dijkstra, E. W. (1984). The threats to computing science. In *ACM South Central Regional Conference*.

Dreyfus, H. L. (2007). Why Heideggerian AI failed and how fixing it would require making it more Heideggerian. *Philosophical Psychology, 20*(2), 247–268. https://doi.org/10.1080/09515080701239510

Harris, Z. S. (1954). Distributional structure. *WORD, 10*(2–3), 146–162. https://doi.org/10.1080/00437956.1954.11659520

Latour, B. (1993). *We have never been modern*. Harvard University Press.

Mikolov, T., Chen, K., Corrado, G., & Dean, J. (2013). Efficient estimation of word representations in vector space. arXiv:1301.3781 [cs].

Rasetti, M. (2020). The new frontiers of AI in the arena of behavioral economics. *Mind & Society, 19*(1), 5–9. https://doi.org/10.1007/s11299-020-00226-4

Russell, S. J., & Norvig, P. (2021). *Artificial intelligence: A modern approach* (4th ed.). Pearson.

Schmandt-Besserat, D. (1996). *How writing came about*. University of Texas Press.

Turing, A. (1950). Computing machinery and intelligence. *Mind, 59*, 433–460.

Vygotsky, L. S. (1987). *The collected works of L.S. Vygotsky* (R. W. Rieber & A. S. Carton, Eds.). Plenum Press.

Wittgenstein, L. (1968). *Philosophical investigations*. Basil Blackwell.

CHAPTER 3

Digital Metaphysics

Ilaria Vigorelli, *Antonio Malo*, *and Giulio Maspero*

Abstract How can a reality whose subsistence is only digital and whose existence is only online have ontological consistency? Yet experience shows how the pervasiveness of communication in human life today and the new possibilities of data analysis, made possible by AI, open up real and crucial ethical questions, in the etymological sense of *crux*. This, in fact, is tied to a "judgement", through the Greek term *krisis* from which the words "crisis" and "critical" descend. A *crux* represents an interpretative passage in which the attribution of meaning is difficult, if not impossible. And this concerns not only philological investigations, but also every "critical" reading of reality.

I. Vigorelli (✉) · A. Malo · G. Maspero
ROR (Relational Ontology Research), PUSC (Pontifical University of the Holy Cross), Rome, Italy
e-mail: vigorelli@pusc.it

A. Malo
e-mail: malo@pusc.it

G. Maspero
e-mail: maspero@pusc.it

M. Bertolaso et al. (eds.), *Digital Humanism*,
https://doi.org/10.1007/978-3-030-97054-3_3

These ethical challenges call into question, at the same time, metaphysics, anthropology and even theology, through the question of what reality can be recognized in virtual relations, and what epistemological criteria, consequently, are required for a conscious manipulation of *big data*. AI and *social media* to actually work and make a profit from human relationships, which are the real product at stake. But to whom does a relationship belong? Who can own symbols and language? The point of access to such questions is the symbolic, anthropological and metaphysical bearing of the relationships themselves.

Keywords Digital humanities · Metaphysics · Artificial intelligence · Big data · Anthropology · Language

1   Introduction: An Oxymoron?

Digital metaphysics is an expression that may appear, from a philosophical perspective, to be an oxymoron: how can a reality whose subsistence is only digital and whose existence is only online have ontological consistency? Yet experience shows how the pervasiveness of communication in human life today and the new possibilities of data analysis, made possible by AI, open up real and crucial ethical questions, in the etymological sense of *crux*. This, in fact, is tied to a "judgement", through the Greek term *krisis* from which the words "crisis" and "critical" descend. A *crux* represents an interpretative passage in which the attribution of meaning is difficult, if not impossible. And this concerns not only philological investigations, but also every "critical" reading of reality.

These ethical challenges call into question, at the same time, metaphysics, anthropology and even theology, through the question of what reality can be recognized in virtual relations, and what epistemological criteria, consequently, are required for a conscious manipulation of *big data*. AI and *social media* to actually work and make a profit from human relationships, which are the real product at stake. But to whom does a relationship belong? Who can own symbols and language? The point of access to such questions is the symbolic, anthropological and metaphysical bearing of the relationships themselves.

The importance of these questions is shown by some concrete examples. Without making any concessions to apocalyptic visions, they already hint at the risk of a communicative totalitarianism that the techno-gnosis in which we are immersed can generate: (1) Cambridge Analytica and psychopolitics; (2) Maatje Benassi as case zero of the Covid-related health crisis in the US; (3) the witch-hunt implemented against Justine Sacco on Twitter (de Souza, 2021). These cases show a decay of humanism, an unavoidable outcome when technology takes over the demand for meaning and freedom, as we are reminded by the "banality of evil" described by Hannah Arendt about the Nazi rise (Arendt, 1963). For this reason, the search for a *digital metaphysics* capable of recognizing the ontological depth of the relationships involved in the open possibilities of IT is essential.

The present contribution is relational not only in its content, but also in its form, because it is the result of the joint research carried out by the three authors within the Relational Ontology Research (ROR) group at the Pontifical University of the Holy Cross.

2 Ethics of the Symbol

A Cartesian approach remains impotent before relations, because the separation between *res cogitans* and *res extensa* cannot grasp the effects of the hybridization taking place between the digital and analogue dimensions (Donati, 2021). The human being is neither only nature, nor only culture, so the change on the communicative form induces an effect also on the content, which is somehow dragged by it. Meanings are always given by the relationship between texts and their contexts. A digital metaphysics, which can support an ethical analysis of the problems posed by the new communicative capabilities, cannot, therefore, be limited to substances, but must also be able to treat relationships as its own objects.

Thus, one of the criteria to evaluate the change underway, which seems to be aiming at the pervasive digitalization of inter-human relations and interactions, not only with things and environments, but also with space–time—i.e., with memory and its objects—is that of the symbolic value of which human life is capable.

By this we mean that dimension not reducible to a concept, but clearly recognizable, which is the ability to signify, that is, to establish relations giving a sense other than the immediate repetition of the acquired datum.

An apparently contradictory effect occurs in human symbolizing: it is the production of a "plus of meaning", ascribable to the experience of the subject and to his creativity, and at the same time of a "minus of communication", which is ascribable to the limits of human knowledge unable to transmit all the being of what it experiences.

The symbolic capacity of the human being is expressed above all in language, in all its forms, from the most initial fragmentary babble of a child who is beginning to speak, to the highest forms of the arts. It always manifests itself as an irreducibility of life to the expressions achieved to represent and transmit it.[1]

Music and literature, cinema and photography, the pictorial and figurative arts, have expressed, along the centuries, a double gap: the one existing between reality and the perception we have of it, but also the one that divides our perception of the world from the ability to represent and communicate it. With Weil we could say that art allows us to call the consciousness of the double gap *attention* (Weil, 1966).

The evocative power of nature and the excess of the world with respect to the code that represents it together provoke human consciousness, inviting it to configure further meanings. This is what happens in poetic language and is what introduces us to wonder, science, metaphysical discourse and religious practice.

These are the configurations that our mind gives to the perception of its own limit, to our impossibility to self-transcend, to the intuition of being finished. A sunset will never be the same after reading *The Little Prince*, but not even Antoine de Saint-Exupéry will have been under the illusion of having been able to find the ultimate expression of the intimate desire aroused in his soul by the poignant sunsets of nostalgia for his beloved far away (de Saint-Exupéry, 1940).

Prudence, the ethical virtue that governs every type of action played out in freedom, once again carries within itself the same irreducibility of the datum to its interpretation, that is, of the choice to the intention that produced it. Ulysses' choice to listen to the sublime song of the sirens cannot be represented on the basis of the calculation of the probability of madness caused by hearing these voices.

We do not understand life, we ascertain it, for, we might say with Kant (*Kritik der Urteilskraft* § 68), one understands perfectly only what one

---

[1] Louis Massignon taught that "one cannot maintain the Memorial of the Encounter if not by entering the night of the symbol", and this is true not only for the communication of Moses' encounter with the Most High, but for every encounter that the human being has with reality. Cf. Campo (2008, p. 131).

can do and accomplish by oneself, according to concepts, and the organization of nature infinitely exceeds all power of a similar representation by means of art.

Attention and responsibility, therefore, are ultimately rooted in the recognition of the symbolic power of the word, in the evocative power of the world and in the excess of human activity in the face of collected data. In a word, we could also call this symbolic power the possibility of love: freedom.

Access to the ethical question on the management of big data—here explored as words on the world—is offered to us by the considerations collected by Plato in the *Phaedrus*, a philosophical dialogue that is right in the foundations of our civilization. Here, the myth of Theuth on the origin of writing is narrated: the Egyptian deity Theuth presented himself to Thamus, king of that region, offering him the gift of writing, a gift that the monarch refused, because it would have induced human beings to remember from the outside and not from within, thus losing the faculty of memory, which is fundamental in metaphysics and Platonic anthropology. It is evident that today we are faced with the myth of Theuth 2.0, because contemporary human beings are offered the possibility not only to externalize their memory with writing, but even to commit it to the cloud by putting it online. The Platonic teaching and its comparison with the communicative technicalism of the Sophists suggests that we should take metaphysically seriously the new passage we are experiencing through the "gift" of the Internet and social media.

In fact, in the original myth Plato introduced the problem of the value of the act of writing as an exemplification of the moral problem connected with the structure of the human soul, i.e., with the problem of the true and the false, the just and the unjust.

A contemporary author such as Jacques Derrida wanted to focus on its perennial topicality as early as the 1970s:

> It is truly morality that is at stake, both in the sense of the opposition between good and evil, or good and bad, and in the sense of mores, public morals and social conventions. It is a question of knowing what is done and what is not done. This moral disquiet is in no way to be distinguished from questions of truth, memory, and dialectics. This latter question, which will quickly be engaged as the question of writing, is closely associated with the morality theme, and indeed develops it by affinity of essence and not by superimposition. (Derrida, 1974, p. 74)

Here, therefore, it is a question of focusing attention on the relational value of the cognitive datum—beginning with the emotions and ending with the technological and digital datum—in its dual capacity of configuring the various representations of reality that we can come to have, and of being a way of accessing the symbolic, that is to say, that narrative capacity which, by overcoming the limits imposed by every type of abstraction/repetition, touches in some way on the inexpressible intimacy of life.

However, it is also a question of overcoming the oblivion of the fact that every linguistic system, even if it can be studied—as Saussure did (de Saussure, 1916)—as an autonomous and unitary system of signs where the functional value of an element is determined by the complex of its relations with all the other elements that make up the system, can never be considered as detached from the other systems, nor be understood in a purely synchronic way, leaving out the diachronic relations relating to the times and contexts that generated it.

What we must try not to forget is that the value of a word (or of a datum) is not provided only by what—reductively—is intended as its meaning, but, from the beginning, by its "use". This refers to the syntactic dimension, which necessarily accompanies the semantic one. When an error is introduced in coding, the computer responds with the expression "syntax error", referring to an inconsistency with the rules of the language being used. The inescapable symbolic dimension of the human being, however, refers here to a larger syntax, which specifically concerns the relationship between the text and the context that lies beyond the language itself. Humanity is played out precisely on this openness to excess, which Plato had already intuited and which the development of AI enhances and highlights.

3   Language, Meaning and Sense

In 1950, Alan Turing, one of the fathers of AI, published his famous article 'Computing Machinery and Intelligence' in the journal *Mind* (Turing, 1950). In it he proposes what will be later known as the Turing test: it is an "imitation game" designed to show how machines can be so intelligent that they can no longer be distinguished from people. The test indicated the necessary conditions so that a judge, finding himself in front of a human person and an intelligent machine (a computer), could no longer distinguish between machine and person, because the machine would have been able to answer to his questions as a person.

One can immediately see that the question, although situated in an AI context, is essentially metaphysical, because it refers back to the question "what is human?", highlighting precisely the need for a digital metaphysics. It must be said from the outset that so far, no computer has actually succeeded in passing this test. But the ethical dilemmas mentioned show that the question itself is unavoidable, together with the indication of a relational criterion to approach it.

In fact, even if computers were able to pass the Turing test, it could not be said that they would have an intelligence similar to the human one. This, for example, is the thesis of John Searle, who, to prove it, devised an experiment similar to the Turing test, known as the "Chinese room" (Searle, 1990).

Instead of a computer, Searle imagines himself, totally unaware of the Chinese language, locked in a room. With the help of an English–Chinese dictionary and all the rules of Chinese grammar, he would be able to answer questions asked by a Chinese person who spoke to him from outside the room, without making a single mistake, so as to induce him to think that he was talking to someone who knows Chinese. Searle concludes his imitation game by stating that, even if he was able to answer and narrate everything he was asked, he would still not be able to understand the meaning of those questions, nor that of his answers and stories.

Searle saw well, in fact, that human intelligence does not consist only in the good use of vocabulary and grammar, that is in the use of formal symbols and in their substitution with other formal symbols, but above all in semantics, that is, in the act of understanding the meaning of words. As a matter of fact, this is the common experience we can have today when we try to use a computer translator, and we realize that our thought is imitated through the words of another language, but not understood and translated. To understand the meaning of a sentence is impossible, if one has no consciousness. If in his experiment Searle succeeds in passing the test, it is because he is able to understand the meaning of the symbols in English and this allows him to correctly use a series of rules thanks to which he can answer questions in Chinese or even ask questions himself, while the computer, since it doesn't understand the meanings, needs a program (software) that allows it to make up for its lack of understanding. An important distinction, therefore, between artificial intelligence and human intelligence consists in the fact that artificial intelligence can use the signifiers (the signs), it can order them, repeat them, combine them according to the rules of a given language, but it is not able to arrive at the meaning of the words, of the rules and of what it is doing in composing sentences, because all this goes beyond the processes of calculation and algorithms", as it belongs to another level, that of the intentionality of consciousness. In this way we are faced with a further gap, that between human intelligence and AI.

However, although this difference is real, it is not the only one, nor the most important one. It would be, in fact, the only one if thought were identified with language (if thought were constituted only by signifier and signified), as it is sustained—as well as by Saussure (de Saussure, 1916, pp. 98–99)—by Wittgenstein and by many analytic philosophers, including Searle himself, because in their perspective, one cannot speak of what goes beyond human language, and therefore not even think it.[2] In other words, the meaning is not only played in terms of semantics, nor can it be reduced to the correctness of the internal syntax of the logical system itself, but refers to a relationship with a beyond.

In fact, human thought completely transcends language, not only *la parole* (that is, the use that the speaker makes of language), but also *la langue* (that is, the set of shared meanings and signifiers that allow speech acts) (de Saussure, 1916, p. 99). The person who knows how to speak Chinese—as we have seen—besides understanding the meaning of ideograms, is in his speaking the bearer of a *plus of* meaning that is not collected by dictionaries, nor in grammar, nor, in the end, in his way of mastering the Chinese language. This *plus* corresponds, for example, to the connotations that words have for her in the sentences she composes. For example, if in your life you have gone through the experience of seeing your house burn down, the ideogram of the house will have special affective connotations for you: nostalgia, sadness, fear…. These connotations, however, can only be found among persons, that is, among beings endowed with an experience whose meaning—or better still, sense—they are capable of discovering.

This capacity for meaningful experience allows, moreover, access to the symbol, not only in a logical but also in a metaphysical sense; thus the house, in addition to having the meaning of dwelling and being endowed

---

[2] "Whereof one cannot speak, thereof one must be silent" (Wittgenstein, 1922, Introduction § 6).

with connotations that reflect experience, can be transformed into the symbol of what is most personal, most proper: a well-known example is that of the dwelling as a symbol of personal intimacy, as observed in the gospel of St. John in which the physical house where the disciple welcomes the Mother of the Saviour is transformed into a spiritual home: in Greek *ta idia*, meaning *his own home* or intimacy (John 19:27).

In this passage to the symbolic we find a further *gap* between human intelligence and AI: the latter, being incapable of meaningful experiences, is consequently incapable of accessing symbols or metaphors, as well as of formulating them when the ordinary language is insufficient to transmit them.

We can maintain, then, that meaningful experience is something transcendental, like goodness, truth, beauty, since it refers to a "being-in-itself" that is also a "being-for-itself". We do not want to propose here a Hegelian vision of human thought,[3] but to open the epistemological reflection to another perspective, relational, in which—in our opinion—we observe the most important difference about artificial intelligence.

The "being-in-itself" is the substance that can be natural—especially the living one—, or artificial, as it is endowed with reality, (this is meant in a broad sense, as is reality, in fact, the imprinted, virtual, digital, etc.). In this sense, the wax tablet, which Aristotle used as an image of human knowledge, resembles the artificial intelligence: both have as reality ("being-in-itself") the capacity of receiving human language, which will be possessed by them, without them also being aware of what they receive. Certainly, AI is capable of processing, calculating, translating, answering questions and also influencing our intelligence to a greater or lesser extent (as happens with search engines and social media), something that the humble Aristotelian tablet does not do. Nevertheless, neither of the two—neither tablet nor AI—are "natural": they are, in fact, artificial, that is, they have been produced to be used, to serve, that is, as an instrument. In this sense, therefore, they are not in themselves an end but a

---

[3] In fact, the in-self for-self corresponds, according to Hegel, to the Spirit as absolute: "In the first situation we had only a *notion* of actual consciousness, the inward emotion, which is not yet real in action and enjoyment. The *second* is this Actualization, as an external express action and enjoyment. With the return out of this stage, however, it is that which has got to know itself as a real and effective consciousness, or that whose *truth* consists *in being in or for itself*" (Hegel1910, italics in the text).

means and, therefore, they cannot "be-for-itself". We are not referring here to a necessary natural law, but to the relational origin of the modern tools provided by IT, which, as products, refer beyond themselves in the sense both of origin and of end. In fact, only living beings are "in-self" a certain end and, therefore, act "for-self" and can arrive—as in the case of the human person—to "be-for-self" in the most proper sense, that is, as free subjects and responsible for their own actions and relationships. This is why human beings are incurable seekers of meaning.

Of course, in comparing artificial intelligence with human thought, we have seen that our thought transcends language and even consciousness because it is capable of experiencing and grasping meaning. Now, we can add that such a capacity depends on its "being-for-itself" in its own way.

All living beings, obviously, live, but not all of them are capable of having experiences (emotions, inclinations, experiences, language): only those who have at least a sensitive consciousness can have them. Such a consciousness, however, which already separates them from AI, is not enough to be able to "be-for-self". One needs to "be-in-relation", as only the person can be, in order to be fully "being-for-self". The relationship of the animal with the environment, in fact, does not become a real relationship because it is not experienced by the animal as a relationship, but it happens only in the experiences to which it gives rise. To live the relationship as such is not even to have the concept of relationship or to be able to use it adequately in certain contexts. Only through the other—the world and, above all, other human beings—is it possible to achieve this goal: we become "being-for-itself" *through the other*. Again, the experiences of the animal originate from relationships with the environment and with other individuals of the species, but do not require other individuals to decipher their meaning, while personal emotions, desires and human language require another human being to reveal their meaning. Therefore, human intelligence not only requires a world and a living being with a consciousness open to the other, but also requires a vital and continuous communication with the other as the one who reveals the meaning of one's experiences and, consequently, the very meaning of existence, which could be formulated as "it is better to be than not to be". From the heart of anthropology emerges, thus, an ontological and relational syntax.

Experience, its expression through language and its communication to others goes beyond the experience of the pure "being-in-itself" that we are, since it contains as an ever-present referent the other as a revealer of meaning. "Being-for-self", then, does not express here the Hegelian reflection of being returning to "being-in-itself" in order to make it fully known (which is impossible because, through reflection, it can never know its origin and destiny), but rather indicates discovering in our existence, in our thought, and in language itself, the presence of the other, which always refers back to the Other as the ultimate source of revealed meaning. The human being is based, therefore, on founding relationships, which go beyond the subject himself and sustain him in being.

The presence of the other is thus found in all human productions and, therefore, also in AI. The intelligence that is required to build computers does not come, in fact, from another machine nor is it autopoietic, but carries within itself, as a mark, the relationality of human existence. However, the machine itself is not relational and therefore it can be used more or less as a tool to improve communication, health, the economic development of all people and not only of the big companies of Silicon Valley. When one loses sight of its being an instrument at the service of the relationship with the other, and not a substitute for that relationship, it then becomes a source of manipulation of the other, that is, a means to make use of the other, reduced to a product from which to obtain the maximum profit, as happens with the commercialization of the profiles of Internet users. This manipulation is evident in the three examples cited in the introduction, where the human and relational dimension is crushed and demeaned by the inexorable and blind necessity of the algorithms.

In this sense Turing was right: it is possible to create algorithms capable of deceiving the human mind, making it believe, for example, that fake news is true or that real news is false, or to manipulate people's behaviour to the point of becoming a danger to their freedom, especially to the extent that it is possible to enter the functioning of the human brain down to its deepest layers, producing in it all kinds of addictions.

However, it is not AI that is the cause of this manipulation and falsehood, since the machine does not know how to distinguish between the true and the false, or between dependence and freedom: the only one responsible for these crimes is the human intelligence that designed it with this intention or at least did not want to take into account the truth and freedom of the people, because what it was looking for was an ever greater monetization of the AI.

On the other hand, the possibility that AI has of producing effects that go beyond the intention with which it was designed implies that human action and its works transcend human intelligence because it, in turn, is only a participation of another intelligence that is infinite.

The referral of human intelligence, with its actions and works, to an infinite intelligence is the cause not so much of the negative phenomena of AI (which are due to the limits of human intelligence) as of the positive ones, such as the perfecting of the human world and the collaboration in the perfecting of people.

This transcendence allows us to speak of an analogy between infinite intelligence, human intelligence and AI, by virtue of which it is possible to evaluate positively or negatively both the use of human intelligence and the use of AI, depending on whether or not the person is taken into account as "being-in-itself and for-self". The answer to the question "what is human?" is not played, therefore, merely at the level of physical nature, but calls into question the free relationality that marks the search for meaning of the human being.

4   Incompleteness and Pandemic Manipulation

When in 2014, Aleksandr Kogan, a professor at Cambridge University, developed an app (called TIDYL = "This is Your Digital Life") to study the personality traits of its users, perhaps no one could have imagined that, thanks to a collaboration that began on June 4, 2014 with the political consulting firm Cambridge Analytica, data mining, data brokerage and data analysis would bring psychopolitics to the world stage for the first time. In fact, not only those who had voluntarily run the personality test on TIDYL, but also their Facebook friends were unknowingly profiled. In this way, the number of those involved had already reached the level of several million in December 2015, when *The Guardian* revealed the collaboration between Cambridge Analytica and Ted Cruz, estimated at 87 million after the revelations of whistle-blower Christopher Wylie in March 2018. This led to a change in Facebook's privacy policy and Zuckerberg's hearing in the US Congress and the European Parliament.[4] The risks to democracy are obvious, as are the ethical questions that data mining opens up. What, then, is the metaphysical value of big data? To

---

[4] This has introduced novelties to the very way journalism is done; see Cf. Venturini and Rogers (2019).

whom does this data belong and, above all, is it possible for people's relationships to become private property, exploited by a company or used to drug a democracy?

The same emerges from the cases of Justine Sacco and Maatje Benassi. The former was a 30-year-old head of communications for a digital media company who, first in New York and then in London while waiting for her flight back to South Africa, posted a message on Twitter joking about her return to Africa and the risks of contracting AIDS. When she put her mobile phone in flight mode before her 11-hour flight, she could not have suspected that, starting with some of her 170 followers, those few dozen characters of hers would be transformed into the number one trending topic by the algorithms that amplify the polarization of opinions, causing her to lose her job and irreparable damage to her reputation, which she only realized once she had landed.

Even more incredible is the fate of Maatje Benassi, a US Army reservist who had attended the Military World Games in Wuhan in October 2019. An American conspiracy theorist posted a video on YouTube pointing to Benassi as case zero of the Covid outbreak and the news was amplified by Chinese algorithms (trolls) to support the narrative of contagion brought from the US to China. In truth, the woman never contracted Covid, though her life was destroyed by this "cyber" epidemic, with no recourse against YouTube.

The moral significance of the action of the algorithms in the three cases above is obvious. But what really happened? What assets were at stake?

The path traced up to now allows us to understand how human relations cannot become objects of commerce and exchange without hurting personal dignity itself and poisoning the communicative habitat. So far, in fact, we have been able to draw a line directing the elements of continuity and those of discontinuity between human intelligence and artificial intelligence, and this leads the reflection back to the first point of our analysis, where we highlighted the double *gap* that characterizes the symbolic depth of human communication: the more of meaning founded in the ontological depth of the subject and a less of communication caused by the expressive limits of knowledge itself.

This double gap can be theologically and philosophically re-read as a trace of the double cognitive and metaphysical dimension of the mystery of the real and, therefore, of the human being. This term—mystery— seems anachronistic in the field of AI and without right of citizenship in the world of IT. The pervasiveness of algorithms, with the promise of automation, seems to leave no room for a mysterious dimension, and, *a fortiori*, for mysticism or contemplation. Everything seems to be revealed or unveilable, everything determined or determinable. It is only a matter of having sufficiently powerful computers, adequately effective programs and, now, a corresponding amount of data.

But the reference to the metaphysical depth of the symbol, which inexorably calls into question the ethical question and responsibility, not only individual but also collective, excludes that this promise of AI can ever be fulfilled. The subject, in fact, as we saw earlier, is right as an end and not only as a means, while the current communicative "market" seems to tend to transform its users into products by taking possession of their relations and their expressive means.

This clash of perspectives has precedents in the history of human thought, which has already learned to define this conflict by the name of "tragedy" for almost 25 centuries. From ancient Greece onwards—and even before, from the clashes between the gods narrated in the foundational myths in different cultures—man has recognized the moments in which his freedom, double-stranded with the purpose that constitutes his being a person, has been wounded by the determinism of irreconcilable laws; think for example of the conflict between obedience to the *polis* and belonging to the *genos*, in Sophocles' *Antigone*. The tragic knot is always connected, in the Greek tradition, to an incomputability, to an irreducibility of man with respect to the formulations that his experience assumes. It is the double gap and the metaphysical depth of the symbol that constitutes the world and the human being to the highest degree. The clash between Plato's metaphysics and the linguistic technique of the Sophists had its roots here.

The greatness of classical humanism consists precisely in this intellectual honesty which led it to clothe with the most sublime beauty the limits encountered by its own thought. These limits have thus revealed themselves to be wounds from which life has sprung: in fact, they have indicated, and will always indicate, the territory that can only be accessed by putting one's freedom and, therefore, one's responsibility on the line, on pain of the lack of meaning that leads to death, to the epidemic that becomes a pandemic.

This awareness seems to be radically opposed to the promise of AI and big-data analysis. Algorithms will anticipate our desires and our health may no longer know danger. Virtual relationships will be the bearer of a secure happiness, because they will be founded not on the uncertainty and imperfection of reality, but on calculation and machine learning. Instead, in the same historical origin of AI was already present the translation of the message of the double gap and the reference to the metaphysical surplus.

Alain Turing, in fact, in conceiving the computer even before it was technically feasible, had taken up and re-expressed the results of Kurt Gödel and his incompleteness theorem (Gödel, 1931). By 1931 Gödel had shattered David Hilbert's claim of algorithmic reduction of mathematical proof. While the research of the time was based on the assumption that it was possible to devise a formal language in which any formulated theorem could be proven true or false by mere computational means, Gödel showed that if such an operation is attempted for a minimal logical-formal system, including trivial arithmetic, the result will be the inevitability of the choice between inconsistency and incompleteness. The starting point of the proof was a reformulation of the paradox of the liar, whose philosophical origin goes back to Parmenides and the Eleatic school (Sorensen, 2005). The expression "this statement is false" was logically translated as "this statement is unprovable". In this way only two logical paths were open: either the assertion could be proved true, but then the system would have given inconsistent results because it could prove as true what is false, or it was not possible to prove either the truth or the falsity of the assertion, concluding that the system was incomplete.

In 1936 Alain Turing demonstrated the non-marginal value of Gödel's proof: he had imagined and formalized a machine able to compute any program without limit of memory, calculation power or time (Herken, 1988). But it was precisely this last element that introduced a criticality, because the infinite availability of time referred to the need to determine the probability that a program would finish, with respect to those that would not stop. Now this "alt-problem", as it is technically defined, is equivalent to Gödel's incompleteness theorem, where the paradoxical statement has become "this statement is not computable" (Turing, 1936, pp. 230–265).

The reformulation in computational terms and the following one, by Gregory Chaitin, in terms of information theory (Chaitin, 1999), show the relevance of incompleteness from the point of view of the question at hand. In fact, if a formal system can never be complete, this means that at some point we will be faced with theorems that cannot be proved true or false from within the system itself. This conclusion leads back to the question of what the logical system wants to represent, that is to say, it introduces the semantic question and the reference to the surplus of an external extra-logical reality, which arises in front of the system itself. In other words, the reference to a syntax that is authentically relational and not merely logical is inescapable.

That the symbolic is inescapable emerges, therefore, not from a position extrinsic to the path that has led to AI, but from its very roots. In fact, every logical-formal system—abstract cognitive structure *par excellence*—is not justified by itself, but refers beyond itself, to the real, invoking the decision of the knowing subject on what the system should represent. This implies the inevitability of judgement: the measurer cannot avoid being measured. Responsibility and the ethical question thus break out, not from the outside, but from within the systems and representations themselves, which cannot close without violating the humanity of those who live in this symbolic habitat.

Expressed in other terms, the symbolic dimension, in its inescapability, refers to the relation, not merely mathematical, but real, metaphysical, as the foundation of the same cognitive possibility in AI. If virtual relations have a truthful content, but are not founded on their own, similarly the analysis of big data can help to grasp the opinions or tastes of a certain portion of the population, but can never respond to the infinite desire that constitutes the bottom of the being of each of those people.

The challenge that today is proposed to *homo technologicus* is more rooted in the *homo* than in the *technologicus*, since, as Turing himself showed, the new computational possibilities are simply incidental to the radicality of the symbolic and, therefore, relational dimension, which indicates the ontological depth of the knowing subject.

We find ourselves on the threshold of an inevitable passage, on pain of tragedy, as the history of thought reminds us. If the ontological depth of the symbol and the relational surplus are denied, we witness a closure of the cognitive apparatus on itself, in which the relationship between means and ends is inverted, in an ineluctably manipulative configuration. In fact, if ontological depth is denied and the otherness of the real is ignored, all technical power is reduced to praxis, i.e., efficiency and performance. In the end, there is a real risk that through virtual relations and big data, human beings will become numbers in their turn, hubs of functional interactions. Manipulation would then be pandemic. What is needed, therefore, is an ecology of the digital cosmos, which allows us to recognize the dimension of freedom within the space managed by algorithms.

5 Conclusion

As a result of what has been said, the double gap and the characteristic of being an end intrinsic to the human being ensure that, even if we crossed the desert of tragedy, this outcome would never be definitive. Of necessity, we will learn to use big data in the service of real relationships, therefore in favour of people and not against them. In fact, every manipulation corresponds to a counter-manipulation that increasingly reduces the space for freedom and responsibility, until it leads to a stalemate where living is no longer possible. Then the critical act becomes inevitable, although it is desirable not to have to go through being nailed to a cross to be able to exercise this judgement.

"Knowledge is power" is an iconic formula of modernity that the new technological possibilities amplify in an unimaginable way, translating it into "algorithm is power". Every closure to symbolic depth and the real relational dimension, however, ends up subverting the relationship between means and ends.

Perhaps the best example of what we would like to express here is what happens to Adam and Eve in the Genesis account of the Garden of Eden in front of the tree of knowledge. It is about the choice that makes them feel fragile and afraid in front of their own nakedness. This reaction is a sign of a limitation that, as long as it was experienced in relationship, did not arouse fear because it indicated the relationship with the creative intellect, experienced above all as a source of life and freedom. If anyone thinks that this theological observation is out of place, it is possible to think of the symbol of the bitten apple that is the logo of one of the world's largest and most innovative IT companies. Will it really be possible to move the human being into a meta-verse, removing us from the symbolic references of the uni-verse, without hurting our humanity?

The observation of the need for a symbolic ecology refers, then, to the need to place at the heart of the digital humanities a digital metaphysics that would make it possible to take care of the relational element in the world of IT and in the process of hybridization that we are experiencing. The symptom that we are mapping, resulting from exposure to this original manipulation, can only refer us, in fact,—through a journey, perhaps a pilgrimage—to a rediscovery of the ontological depth of the symbol and the mystery; we could then say, not without risk, that the symptom shows us the originating relationship that founds the excess of the real and the double gap in the exceeding relationship with the Father, with our own being children, not produced, nor computed, but generated.

REFERENCES

Arendt, H. (1963). *Eichmann in Jerusalem: A report on the banality of evil.* Viking Press.

Campo, C. (2008). *Gli imperdonabili.* Adelphi.

Chaitin, G. J. (1999). *The unknowable.* Springer and Id.

de Saussure, F. (1916). *Cours de linguistique générale* (C. Bally, A. Sechehave (Eds.) with the collaboration of A. Riedlinger). Payot.

de Saint-Exupéry, A. (1940). *Le Petit Prince.* Gallimard.

Derrida, J. (1974). *Dissemination.* The Athlone Press.

Donati, P. (2021). *Transcending modernity with relational thinking.* Routledge.

Gödel, K. (1931). On formally undecidable propositions of Principia Mathematica and related systems I. *Monatshefte fur Mathematik und Physik, 38,* 173–198.

Hegel, G. W. F. (1910). *The phenomenology of mind* (J. B. Baillie, Trans.). Swan Sonnenschein & Co. (Original work published 1807)

Herken, R. (1988). *The universal turing machine.* Oxford University Press.

Searle, J. R. (1990). Is the brain's mind a computer program? *Scientific American, 262*(1), 25–31. http://beisecker.faculty.unlv.edu/Courses/PHIL%20330/Searle,%20Is%20the%20Brain's%20Mind%20a%20Computer%20Program.pdf

Sorensen, R. (2005). *A brief history of the paradox: Philosophy and the labyrinths of the mind.* Oxford University Press.

Souza, K. G. (2021). *Mapping out moral dilemmas of free expression on social media. A closer look at Facebook, Twitter and YouTube* (Doctoral thesis). Pontifical University of the Holy Cross, Rome, pro manuscripto.

Turing, A. (1950). Computing machinery and intelligence. *Mind, 59,* 433–460. https://phil415.pbworks.com/f/TuringComputing.pdf

Turing, A. M. (1936). On computable numbers, with an application to the Entscheidungsproblem. *Proceedings of the London Mathematical Society, 2*(42), 230–265.

Venturini, T., & Rogers, R. (2019). "API-based research" or how can digital sociology and journalism studies learn from the Facebook and Cambridge Analytica data breach. *Digital Journalism, 7,* 532–540.

Weil, S. (1966). *Attente de Dieu.* Fayard.

Wittgenstein, L. (1922). *Tractatus Logico-philosophicus* (C. K. Ogden, Trans.). Routledge & Kegan Paul. (Original work published 1921)

CHAPTER 4

The Data of the Rose

Roger Strand◉ *and Zora Kovacic*◉

Abstract Information and communication technologies (ICTs) are changing us and our worlds in subtle, sophisticated and sometimes violent ways, as much as any other of the technologies that rely on science. Today's Romeos and Juliets are checking their phones rather than smelling the roses or listening for birds. And so they won't notice that our ecosystems are approaching their tipping points and the spring has become silent.

Keyword ICT · Digitization · Complexity · Alienation · Symbolic violence

R. Strand (✉)
Centre for Cancer Biomarkers (CCBIO), Centre for the Study of the Sciences and the Humanities, University of Bergen, Bergen, Norway
e-mail: roger.strand@uib.no

Z. Kovacic
Universitat Oberta de Catalunya, Barcelona, Spain
e-mail: zkovacic@uoc.edu

Centre for the Study of the Sciences and the Humanities, University of Bergen, Bergen, Norway

M. Bertolaso et al. (eds.), *Digital Humanism*, https://doi.org/10.1007/978-3-030-97054-3_4

49

1   Knowledge is Power

'Tis but thy name that is my enemy;
Thou art thyself, though not a Montague.
What's Montague? It is nor hand, nor foot,
Nor arm, nor face, nor any other part
Belonging to a man. O, be some other name!
What's in a name? That which we call a rose
By any other name would smell as sweet;
So Romeo would, were he not Romeo call'd,
Retain that dear perfection which he owes
Without that title. Romeo, doff thy name,
And for that name which is no part of thee
Take all myself.

Western thought and European culture have for centuries, indeed for millennia, been plagued by dichotomies. With them, we—that privileged historical subject—have sought to purify and separate into categories of humans and animals, men and women, citizens and slaves and, more fundamentally, *we and the others*, sometimes to our own detriment but more than often to the detriment and destruction of those others.

Dichotomies of the type exemplified above served to populate a world-view whose technological realization included little more than naming, identification and blunt violence. Other dichotomies, often associated with the contributions to Western thought made by the philosophers of Ancient Greece, became the cornerstones of what in the course of history developed into European science and modern technology. These immensely powerful dichotomies purified and separated more subtle cate-gories, such as mind and body; the knowing immaterial subject and its knowable material objects; sign and referent; information and matter; and knowledge and action. As vividly portrayed in Bruno Latour's seminal work *We Have Never Been Modern* (1993), armed with this purifying epistemological discourse, European and later global science managed to create a modernity with technologies that allowed limitless ontological transgressions of the categories, to the point that the atmosphere and even parts of outer space are now littered with human produce, while humans themselves devote ever more of their lives to so-called virtual worlds and spaces.

Abusing their slaves and killing barbarians, the Roman Empire never exceeded a human population of 100 million. As elegant horse-riders and efficient butchers, the Mongols created an empire that became slightly larger. In 2021, at the time of writing of this text, however, the human population is approaching 8 billion individuals who account for 36% of the biomass of all mammals. Cows, pigs and other livestock account for 60% of the mammals, while wild mammals have a 4% share, in part because we have outnumbered them but mostly because we have killed and displaced them. As for birds, 60 years after Rachel Carson's *Silent Spring*, poultry vastly outnumber wild birds. Wilderness is largely eradicated, ecosystems are largely disrupted and in the vicinity of irreversible tipping points, as are water and nutrient cycles, the climate system and energy supply systems. In order to fill the Earth, humanity has wasted it.

2 Urban Delirium

This precarious state of affairs is well documented by science and acknowledged by most governments. Still, or perhaps precisely *because* of the gravity of the situation, human desires and human creativity ever more direct themselves away from the material world and into the so-called virtual one, of the internet and its promises and pleasures. While birds and bees are disappearing, we develop ICTs to collect and manipulate more data about them. While rational public discourse is replaced by manipulation in social media or brute political force, creativity is invested in developing artificial intelligence in place of human reasoning. Transhumanists, the vanguard of this movement away from the material and towards the virtual world, see the promise of Paradise in the digital revolution: soon, they hope to reach the Singularity, when the contents of their minds can be uploaded to the data cloud, and they can dwell forever as knowing immaterial subjects in a digital Heaven.

Governments that resist the transhumanist daydream and try to be responsible postulate another type of technological miracle, by which more of the same epistemological, ontological and technological approaches that brought humanity to its tipping point, somehow will save the day and solve our environmental and existential problems. Somehow, more collection of data about humans, bees, roses, winds and soils, and more advanced machine learning to process the resulting libraries with zeta- and yottabytes, will make it possible to govern people and ecosystems and achieve a sustainable future, as if the problem was one of inefficient logistics.

Such beliefs have been discussed from the perspective of psychoanalysis and even diagnosed as delirious (Giampietro, 2018). There might be many ways out of the delirium. As an attempt at reverse psychology, one could ask: What would a truly artificial and truly intelligent machine recommend? If it knew ecology and had the data—and was artificial enough to remain emotionally unaffected by the current human predicament—it might point to the simplest solution of them all: the populations of humans and their livestock are too large and have to be reduced a lot, and sooner rather than later. A similar thought experiment inspired the 2004 movie *I, Robot*, in which robots have to obey the law of not allowing humans to come to harm. As artificial intelligence develops, it comes to the conclusion that humans harm themselves by destroying their environment and the resources they depend on, and thus machines paradoxically turn against humans to prevent humanity from harm.

3   THE ROSE THAT CAN BE
NAMED IS NOT THE REAL ROSE

In this chapter, we take a different route of epistemological therapy, namely that of revisiting "the original sin", the conceptual dichotomies that enabled modern science and technology. Some of these dichotomies played an important role in their time. The dichotomy between truth and belief made it possible for modern science to emerge as something different than religion. Dichotomies create a binary vision, simple categories that can provide useful heuristics in certain contexts. Binary thinking, on the other hand, is not the best tool to navigate complexity and uncertainty. Nevertheless, dichotomies are eminently reified in ICTs and in the current worship of data and information as means of tackling the complexity of the world. The digital transition means translating the world, its complexities and differences, into a binary code, where things can be either a 0 or a 1. Importantly also, the digital transition makes it possible to situate the complexity "out there", to render it as a characteristic of the material world to be tamed by data.

More fundamentally, however, the conceptual dichotomies of Western thought are all known to be *simplifications of the world*. The mind and the body are entangled and not separate entities. The knowing subject is not immaterial and the known and named object is affected by the knowing and naming. The relationship between the signifier and the signified is neither simple, wholly arbitrary nor fixed. Complexity is very much an epistemological predicament, it is the everlasting challenge of making sense of contradictions, inconsistencies and change—within and outside of ourselves. Romeo is a Montague with everything that entails; should he "doff his name", he might indeed enjoy some freedom to become somebody else than he once was, especially because he was conceived well before the era of governmental biometric databases. And at the same time, even if Wikipedia pretends to know that information…

> … answers the question of 'What an entity is' and thus defines both its essence and the nature of its characteristics,

the information about Romeo is not Romeo, as little as the data of the rose can smell so sweet.

Who is Romeo? On one hand, he is a Montague, the son of Lord Montague in Shakespeare's play and as such an echo and interpretation of the "real" Montagues or Montecchis (Moore, 1930) who were an Italian thirteenth-century faction and not a family. In the play, though, Romeo is a family member, not of just any family but one of status and in constant feud with the Capulets. And more than that, he is Romeo, the beloved son of his mother, the young man who already before he meets Juliet, expresses himself as a man of passion, and he is being recognized as such. Born to be free, he takes and receives his role as the passionate one, and so he is made into and becomes that person. However, as the protagonist of Shakespeare's tragedy, a play republished, re-edited and reinterpreted ever since its creation, he is all of this and much more. Every new generation of male actors find new facets of Romeo to emphasize, and more will come. There can be no exhaustive list, no complete inventory, no comprehensive data for what and who Romeo will be in the minds of present and future actors, spectators and readers.

And this is so even if Romeo only exists as fiction. Romeo's richness and complexity is due to the author and what he might have heard about the Montecchis, but above all due to the richness and complexity of every human that interprets and engages with this text. In a world filled with constraints—laws, structures, symbolic and physical violence and discipline—human engagement with art and literature is one of the remaining lucid expressions of our innate ability to be free, not only to react within the constraints but even to act upon them and change them and by that, change ourselves. This is one way we can understand Lao-Tzu's dictum that "the Tao that can be named is not the real Tao". Our ways of acting and being are more than what can be described. We are not a simple form.

Lao-Tzu's claim was general and did not merely refer to the category of humans. It was a combined ontological and epistemological claim about the world as such and our capacity to know it; one of its corollaries would be that the rose that can be named is not the real rose. For instance, the real rose is a biological organism, an individual but at the same time a species (or indeed many of them), a *holon* in the sense of philosophy of biology that exists across time both as individuals and an evolving process of becoming. The real rose is the biological predecessor of offspring that in due time will have evolved to something else than roses, in close interaction with their habitats and ecosystems. And nowadays, a major component of the ecosystem of the rose is the human species that cultivates it and allows it to take part in our most sacred moments, as when today's Romeos approach their Juliets; when they possibly marry; even in their funerals. In Latour's word, the rose is an *actant*. It is a hybrid between nature and culture to be found in myriads of networks between humans and non-humans; and it does its work in these networks with formidable agency. Norway, the home country of one of the authors of this chapter, saw a dark moment in 2011, after the hideous acts of a xenophobic and disturbed terrorist. The immediate response of the Norwegian population was to fetch roses. Across the country, there were mass demonstrations, quiet, peaceful and in mourning, with hundreds of thousands of roses brought to the streets, the squares and the churchyards. Who else than roses could be powerful enough to fight this battle against hatred?

4   ADDICTED TO THE STUDY OF CARTOGRAPHY

The idea of information belongs to many things, including an Aristotelian worldview. Still, the richness of the Aristotelian thought transcends that of modern science. Aristotelian accounts would need to consider not only the formal cause but also the material, efficient and final causes of whatever is to be accounted for. Modern science became a success by temporarily bracketing the final causes, and abstracting away the material ones. The Platonic vein in modern science, the idealism so well articulated by Galileo, led to the programme of obtaining control over dead matter by first manipulating it within a world of ideas, in shapes such as variables, numbers and equations. This works well when systems are closed or otherwise controlled, under "ideal" conditions, as it is called, without friction, contamination or other inference from the real world. The universal computer, the Turing machine, can be considered the apex of this programme: it can simulate everything, it was claimed. And so our civilization—that of twentieth- and twenty first-century modern science—came to believe that climate change, the stock market, fish stocks, earthquakes and criminal behaviour can be predicted and hence controlled by a Turing machine, if it is only is made bigger, with more data and more sophisticated algorithms.

Now, it cannot, not really, not even numerically, and this has been known for long, since 1963, when Edward Lorenz discovered the butterfly effect. Indeed, in a sense it was known already by Henri Poincaré more than 100 years ago and anticipated in the middle of the nineteenth century by Augustin-Louis Cauchy, who took utmost care in defining infinitesimals and thereby stating the highly demanding conditions to be satisfied for calculus to be valid and reliable. Later, Robert Rosen (1991) discussed the causal structure of real organisms and not just the abstract butterfly of Lorenz's thought experiment. In Rosen's analysis, the existence of final causes—*functions*—in biological organisms implies that they have an entailment structure that cannot be brought into isomorphism with mechanisms and differential equations. Even if we accept the Aristotelian scheme, the formal cause alone is not the essence.

Rosen explained the dimension of complexity that is already there in the single biological individual. When we also take the phylogenetic dimension of the holon into account as well as its existence in time, in evolutionary processes in ecosystems, we are reminded that final causes also evolve. If we were to provide an account of all the potentialities of the rose, we would indeed end up offering the rose itself, as Jorge Luis Borges so eloquently explained in his short story "On Rigour in Science". Humans and roses are part of an open-ended evolutionary process, and this is why the megamodel with the yottabytes cannot solve our contemporary environmental and social challenges. They are not problems of inefficient logistics.

5   CROSSING THE STYX

Or are they? More precisely, can they be made into problems of efficiency and control? As Isabelle Stengers (1997) remarked, science can have one thousand and one sexes, and the difference between stones and humans is that humans are so versatile and flexible. We seem to be able to accommodate any map by changing our own external and internal terrain. For Hannah Arendt (1973), this was the deepest cause of the Holocaust: that for humans, existence does precede essence and we can be told to do anything, including marching into the gas chambers. Physical violence is often enough; symbolic violence is no less powerful. If the forest is not isomorphic to the Turing machine, make it into a monocultured field. If human behaviour escapes prediction, catch it again, train and discipline it to correspond to algorithm. The question is then no longer an epistemological one, if Rosen was right and biology has to adopt non-mechanistic models. It becomes one of technoscience and phenomenotechnique: how do we engineer life, from the nanometre to the kilometre scale, so that its open-ended material and semiotic features no longer become prominent? In the words of Strand and Chu (2022), how can we cross the Styx?

It is in this light that the future of ICTs seems so bright. Armed with the concepts of "data" and "information", human proponents of ICT's imaginaries can perform the discursive work of purification needed to convince and recruit their fellow humans: nothing special going on here, nothing to worry about, we are just describing the world even better, for our and your own good. The proponents may even believe in these claims for a little while.

On the very mundane and practical level, let us recall that ICTs are material objects that humans use to store and manipulate information. These objects have to be made from natural resources (including rare elements), take up space and consume energy. For instance, in 2020, bitcoin was reported to consume more electricity than Argentina. So far, the environmental toll of ICTs and the quest for big data and fast computation has been offset thanks to Moore's law. At the end of the day, however, transistors cannot be smaller than a silicon atom. Moore's law will cease to apply in less than a generation.

Humans have been trying to engineer life for a long time, in many different ways. We may recall Maturana and Varela's (1972) work on autopoiesis, and their attempt to write the algorithm of life, an algorithm which was encoded to keep creating itself. The algorithm, however, failed to evolve and was only "alive" for a little while. Experiments like this and efforts to create an artificial intelligence have spurred interesting debates about what is life and what is intelligence. At the same time, artificial intelligence redefines human intelligence, and as the new generations are learning from a very young age to use digital devices and operate in the digital world, new types of intelligences and intellectual skills are being created by and with ICTs.

And so, can the ICTs perform their work of hybridization, of connecting and entangling themselves ever tighter onto the lifeworlds of humans and other living beings, imposing constraints on behaviours and thoughts, and training us to become more like them, to respond and fit better with their types of agency? They will change us and our worlds in subtle, sophisticated and sometimes violent ways, as much as any other of the technologies that rely on science rather than the Mongol art of butchering, as when today's Romeos and Juliets are checking their phones rather than smelling the roses or listening for birds. And so they won't notice that the spring has become silent.

References

Arendt, H. (1973). *The origins of totalitarianism*. Harcourt Brace Jovanovich.

Giampietro, M. (2018). Anticipation in agriculture. In R. Poli (Ed.), *Handbook of anticipation*. Springer. https://doi.org/10.1007/978-3-319-31737-3_23-1

Lao-Tzu. (~6th century BC/1993). Tao Te Ching. Hackett Publishing Company.

Latour, B. (1993). *We have never been modern*. Harvard University Press.

Maturana, H. R., & Varela, F. J. (1972). *Autopoiesis and cognition: The realization of the living*. Boston studies in the philosophy and history of science. Reidel.

Moore, O. H. (1930). The origins of the legend of Romeo and Juliet in Italy. *Speculum, 5*, 264–277. https://doi.org/10.2307/2848744

Rosen, R. (1991). *Life itself*. Columbia University Press.

Stengers, I. (1997). *Power and invention: Situating science*. University of Minnesota Press.

Strand, R., & Chu, D. (2022). Crossing the Styx: If precision medicine were to become exact science. In A. Bremer & R. Strand (Eds.), *Precision oncology and cancer biomarkers: Issues at stake and matters of concern* (pp. 133–154). Springer. https://doi.org/10.1007/978-3-030-92612-0_9

PART II

Computability of Human Problems

CHAPTER 5

Experience-Driven Decision Support Systems: The Analytic Hierarchy Process

Martina Nobili, Gabriele Oliva, and Roberto Setola

Abstract Automatic decision support systems are typically based on objective data and rely on data-driven techniques such as machine learning. Yet, in order to take effective decisions, it is fundamental to incorporate also *experience-driven* approaches that are able to leverage on the experience of human decision makers and experts. However, there is a need to handle the inconsistencies, contradictions and subjectivity that are typical when human actors are involved. This chapter reviews the analytic hierarchy process technique as a tool to derive absolute relevance values based on relative preference information provided by human decision makers. The main ideas of the methodology are informally presented,

M. Nobili (✉) · G. Oliva · R. Setola
University Campus Bio-Medico of Rome, Rome, Italy
e-mail: m.nobili@unicampus.it

G. Oliva
e-mail: g.oliva@unicampus.it

R. Setola
e-mail: r.setola@unicampus.it

M. Bertolaso et al. (eds.), *Digital Humanism*,
https://doi.org/10.1007/978-3-030-97054-3_5

61

a suite of different possible applications is reviewed, and considerations regarding the relation between human and automatic decision-making are discussed. The chapter is complemented by an Appendix providing a formal mathematical presentation of the key concepts.

Keywords Decision support systems · Multi-criteria decision aiding · Analytic hierarchy process · Experience-driven systems

1   Introduction

Decision theory is becoming a fundamental pillar in the development of innovative, automatic ways to support humans in taking effective decisions, with applications ranging from medicine to cyber-security and from economics to climate change. Notably, in recent years, machine-learning techniques have proved a remarkably useful tool to filter and aggregate huge streams of data, but in order to be effective in real-world decision scenarios they need to be complemented by the experience of human actors, decision makers and subject matter experts. Indeed, in several contexts, SMEs hold valuable experience and knowledge which, if suitably incorporated into an automatic decision framework, can significantly improve a mere data-driven approach. Among other techniques in the literature, in this chapter we consider the analytic hierarchy process (AHP) formalism (Saaty, 1977) (see Golden et al., 1989; Ho, 2008; Saaty & Vargas, 2012; Subramanian & Ramanathan, 2012; Vaidya & Kumar, 2006 for a wide variety of applications) in both a complete and incomplete information setting, where SMEs are asked to provide comparisons between pairs of alternatives (e.g., "the alternative X is twice as important as the alternative Y"), and such relative data is elaborated in order to derive an absolute degree of preference for each alternative. The AHP methodology enables subjective information to be incorporated into a systematic mathematical model; it can be adopted to combine a wide variety of conflicting metrics and indicators appropriately, reconciling them in a way that reflects the SMEs' preferences. Note that the output of the AHP methodology is characterized by meta-information regarding the quality of the estimated absolute utilities that are based on the subjective opinions of an SME; this allows the AHP formalism to be adopted as a framework to identify opinion groups in

SMEs, with applications ranging from marketing to political analysis. The AHP methodology represents a paradigmatic framework for discussing the relationship between human decision and automated decision, and identifying ways to reconcile these orthogonal domains.

The chapter follows the following conceptual outline. First, the main ideas behind the AHP in both complete and incomplete information settings are presented. The main ideas of the methodology are explained, using minimal mathematical notation to reach a broad and non-technical audience. Applications of the AHP framework are discussed, with particular reference to the identification of opinion groups in SMEs, and the relation between human and automatic decision-making is analyzed with particular reference to the AHP methodology. The chapter is complemented by an Appendix providing a formal mathematical presentation of the key concepts presented, in order to allow the interested reader to apply the methodology to real-world problems.

2   Main Ideas Underlying the Analytic Hierarchy Process

Let us imagine ourselves in Bologna, Italy. A passer-by asks us: "How tall is the Garisenda Tower? What about the Asinelli Tower?" Can we confidently provide an answer? Figure 1 is a view of the two towers from the ground (Garisenda, left; Asinelli, right). While everyone would easily agree that the Asinelli is taller than the Garisenda, coming up with an exact value seems difficult.[1]

This example represents an instance of the dichotomy existing between absolute and relative measurements. Very often, we are in a situation where we would like to assess the absolute measure, relevance or importance of one or more phenomena or alternatives, but an "expert" or "decision maker" is only able to provide a relative estimate, which is typically of ordinal (e.g., "The Asinelli Tower is taller than the Garisenda Tower") or cardinal (e.g., "The Asinelli Tower is *twice* as tall as the Garisenda Tower") nature.

Is it still possible to assess an absolute value, relevance or importance of the alternatives and to rank them, in spite of the relative nature of the available information? Notably, if we are given the above piece of

---

[1] The Garisenda Tower is 48 metres high and the Asinelli Tower 97 metres high.

Fig. 1 The two principal towers of Bologna, Italy: the Garisenda Tower (left) and the Asinelli Tower (right) (*Source* https://it.wikipedia.org/wiki/File:Bologn a,_2_Torri.jpg)

cardinal information regarding the two towers, we can conclude that if the Garisenda Tower is 50 metres high, then the Asinelli Tower is 2 × 50 = 100 metres high. This example demonstrates that, based on relative information, it is possible to obtain an absolute value only if an additional reference value is available (e.g., we could have assumed 75 metres for the height of the Garisenda Tower, thus obtaining 150 metres as an estimate of the height of the Asinelli Tower). In other words, rather than "absolute" values, here we seek "intrinsic" values that can be associated to each alterative, even though such values are defined up to a scaling factor; yet, such values could effectively be used to rank the considered alternatives.

Let us now consider a scenario where there is a need to rank three alternatives A, B and C (e.g., three towers, based on their height). In this case, one may expect to collect *consistent* relative preference information as in the case of the left panel in Fig. 2. Yet, a typical scenario involving human decision makers is more similar to the case depicted in the right panel. In the latter case, Tower A is twice as tall as Tower B and Tower B is four times as tall as Tower C, but Tower A is only seven times taller than Tower C: this represents an *inconsistency* (i.e., we would expect A to be 2 × 4 = 8 times taller than C).

Indeed, quite often, human judgements are contradictory or inconsistent (Hill, 1953; Coombs, 1958; Regenwetter & Davis-Stober, 2012); this is especially true when our desires and beliefs are involved (Hoch &

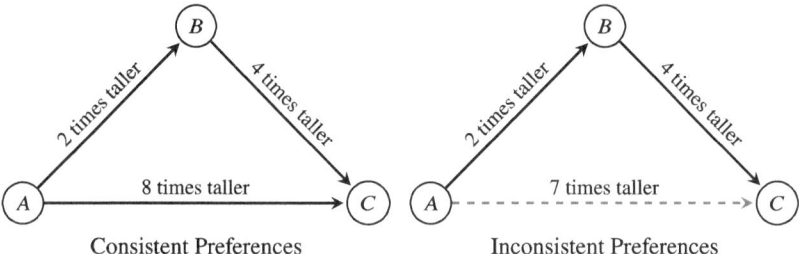

Fig. 2 Consistent and inconsistent relative preference information

Loewenstein, 1991). Figure 3 shows an example of relative preferences regarding the political leaders running for election in Italy on March 4, 2018 (Oliva et al., 2019). Notably, the respondent whose opinion is depicted in the figure liked Matteo Renzi four times more than Silvio Berlusconi and Silvio Berlusconi twice as much as Matteo Salvini, but preferred Matteo Renzi six times more than Matteo Salvini (and not eight times as one would expect in a consistent case). Moreover, it is interesting to observe that the respondent did not compare some of the alternatives (i.e., Grasso, Bonino), thus implying that the decision maker did not have particular preferences for a subset of the alternatives: this *incompleteness* is a feature that is typical of human decision makers. Several approaches have been developed, despite incomplete information, to obtain absolute preferences (e.g., see Bozóki et al., 2010; Menci et al., 2018; Oliva et al., 2017).

Let us now consider a simple but illustrative example. Figure 4 shows the relative preference of a decision maker regarding four different sweets. Let us define w_1, w_2, w_3 and w_4 as the intrinsic utility (defined up to a scaling factor) for the four sweets: *cannoli, babá, montblanc* and *pie*, respectively. Our aim is to find a numerical value for w_1, w_2, w_3 and w_4 which is as concordant as possible with the available information. Interestingly, in the example, *cannoli* is liked twice as much as *babá* and three times more than *montblanc*, but *babá* and *montblanc* are liked the same. In this inconsistent case there is no choice for w_1, w_2, w_3 and w_4 that perfectly matches with the relative preferences. Therefore, a reasonable way to choose the absolute preferences is to seek values such that, at the same time, it holds $\frac{w_1}{w_2} \approx 2$, $\frac{w_1}{w_3} \approx 3$ and $\frac{w_2}{w_3} \approx 1$; notably, we will need to reach a compromise or equilibrium among such conflicting objectives.

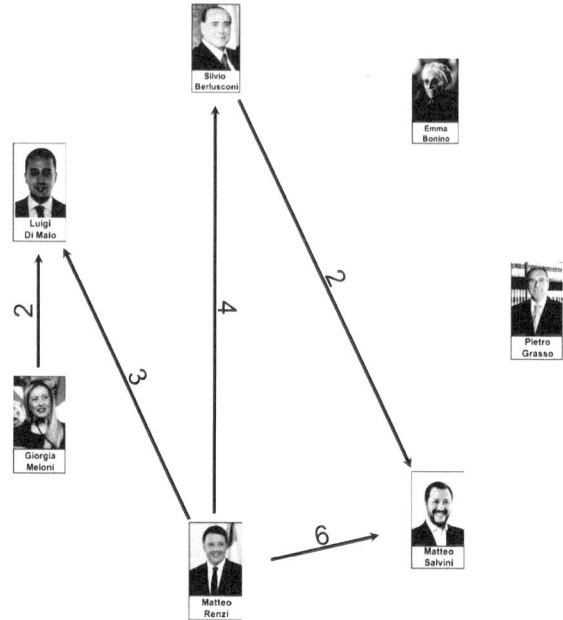

Fig. 3 Sample relative preferences for the political leaders participating in the Italian election on March 4, 2018 (The figure is based on the data available in Oliva et al. [2018]. The pictures of the leaders are under Creative Commons 2.0 license or are free to use, and are retrieved from https://commons.wikime dia.org)

Several methods can be used to obtain the approximated values. Although to review these is beyond the scope of this chapter, the Appendix contains a detailed description of the *logarithmic least-squares* approach (LLS), an optimization technique that attempts to find a good compromise and can be easily computed in practical cases. In particular, the LLS approach applied to the example in Fig. 4 would yield values $w_1 = 7.86$, $w_2 = 3$, $w_3 = 3.43$ and $w_4 = 1$, which represent a good compromise (actually, the best possible compromise according to a criterion that will be formally described in the Appendix), i.e., $w_1/w_2 = 2.62$ (and is thus close to 3), $w_1/w_3 = 2.29$ (close to 2) and $w_2/w_3 = 0.87$ (close to 1). Moreover, the solution of the problem is associated with a numerical value (i.e., the value of the *objective function*), which represents a measure of the overall

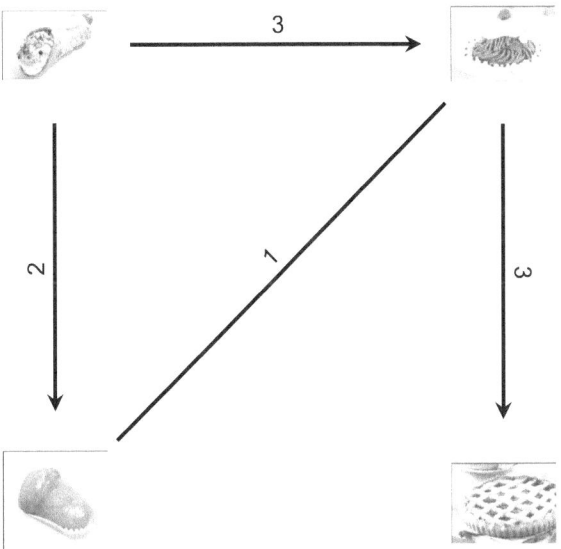

Fig. 4 Example $w = [7.86 \ 3 \ 3.41 \ 1]^{\mathrm{T}}$ (#1) *Cannoli* (#2) *Montblanc* (#3) *Babá* (#4) Pie

degree of inconsistency (zero if perfectly consistent, larger than zero and proportional to the degree of inconsistency otherwise); in this view, this measure can be regarded as a *meta-information* regarding the quality of the available relative preferences. For the example at hand, LLS yields a value of 0.1634, which can be considered as small, implying that the available data is characterized by a small degree of inconsistency.

3 Selected Applications

This section reviews three applications of the above approach that demonstrate its utility and effectiveness in different contexts.

Finding Opinion Clusters

In Oliva et al. (2019), we used the LLS approach to incomplete AHP to identify groups of decision makers holding similar opinions. Consider a scenario where a group of decision makers provide partial relative preference information regarding a set of alternatives. The main idea is to divide the decision makers into groups such that, when the relative preferences of the decision makers in the group are combined and a group absolute preference is computed (as shown in the Appendix), the inconsistency associated with the group is small. In Oliva et al. (2019), a novel clustering algorithm based on LLS and AHP is developed for this purpose. In other words, rather than being interested in the actual preferences of the single decision maker, we seek groups that are characterized by similar opinions and a small degree of overall inconsistency. Considering the political voting scenario represented in Fig. 3, we experimentally observed that, when the decision makers were clustered into three groups, there was a significant drop in group inconsistency, suggesting that the groups featured decision makers with compatible opinions. When the three groups were reviewed (see Fig. 5), we obtained small degrees of inconsistencies. Moreover, by analyzing the absolute preferences in each group, we discovered that the groups reflected the political tendencies existing at the time of the election, i.e., we obtained a group with a preference for the left-wing coalition (red), one group preferring the right-wing coalition (blue) and one group with a large preference for the Movimento 5 Stelle party (yellow), who ran as an outsider. Overall, this example demonstrates the versatility and applicability of such methods for identifying opinion clusters. Possible further applications include, among others, product planning, voting or social media analysis and consumer segmentation.

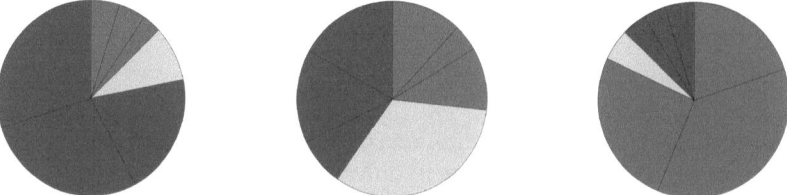

Fig. 5 Three opinion groups obtained by clustering decision makers based on their relative preference on political leaders (immediately after the 2018 Italian elections) (Oliva et al., 2019). The colour red is used for political leaders in the right-wing coalition, blue for the right-wing candidates and yellow for the Movimento 5 Stelle party

Planning for Construction Site Locations in Large Infrastructures

With large infrastructure construction sites increasingly becoming the target of violent protests by radical environmentalist and ideological groups, careful selection of sites among possible alternatives is needed, bearing in mind the risk associated with violent action. However, in many cases one cannot resort to the traditional definition of risk as the product of probability by impact. While the impact might be assessed with sufficient accuracy, the probability of an attack is hard to ascertain (e.g., due to the lack of historical and/or reliable data). In Oliva et al. (2021), we suggested overcoming this limitation by considering risk as the product of *attractiveness* by impact, where by attractiveness we mean the appeal of a construction site. Such an abstract and faceted index is developed in Oliva et al. (2021), based on the combination of fourteen criteria, including aspects such as the expected damage, the symbolic value of the construction site, the environmental and morphological characteristics of the site, the risk of collateral damage (the latter aspect being relevant for protesters seeking popular support). From a technical standpoint, the attractiveness indicator was constructed based on the relative opinions of subject matter experts, which were combined via LLS (as explained in the Appendix). The indicator was applied to the analysis of the potential construction sites for the Turin–Lyon high-speed and high-capacity railway. Figure 6a shows the assessment of impact and attractiveness associated with the ten construction sites that characterized the original construction plan, while Fig. 6b depicts the best solution found, which features eleven construction sites characterized by an overall reduced risk (e.g., only one site belongs to the red zone with large associated risk).

Combining Different Metrics to Identify Important Subsystems in Infrastructure Networks

The task of identifying critical subsystems in infrastructure networks (e.g., elements that require more protection or attention) is fundamental in order to implement adequate protection strategies. Yet, in the literature there are several, often conflicting, definitions of the importance of an element in a network, each capturing a particular facet (e.g., a node is important if it has many connections, if its removal would disrupt the connectivity, if it is "central"; see Rodrigues (2019) for a discussion on different centrality and importance measures). In Faramondi

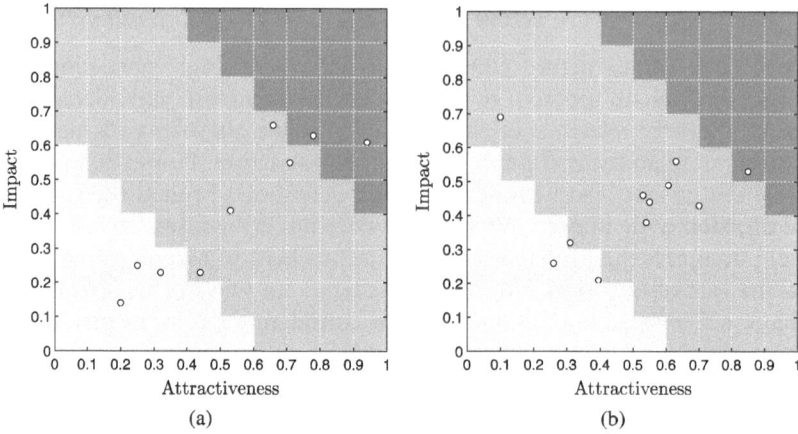

Fig. 6 Panel (a): results of the assessment for the ten sites that constitute the nominal configuration (reported by black circles with white interior). Panel (b): results of the risk assessment for the optimal configuration, involving eleven sites (reported by black circles with white interior) (*Source* Oliva et al., 2021)

et al. (2020), we develop a holistic indicator of the importance of nodes in a network that combines several centrality or importance measures. Notably, the holistic indicator amounts to a weighted sum of the (normalized) centrality measures, where the weights are derived based on LLS AHP based on the relative preferences of experts. Specifically, we consider as a case study the central London tube network (reported in Fig. 7), considering the stations as nodes in the network and the existence of a direct path among stations as a link. Twelve different metrics are combined (Fig. 8), with each index identifying different important nodes (shown as yellow and orange).

Figure 9 shows the importance of the nodes according to the proposed holistic indicator; the proposed methodology represents a good trade-off between the different metrics, as it assigns large importance to the elements that are most influential according to the single metrics. Moreover, in Faramondi et al. (2020) we experimentally demonstrated that the ranking obtained via the proposed index shows no correlation with any of the rankings obtained based on the single metrics, suggesting that the proposed approach does indeed provide new information. Overall, the

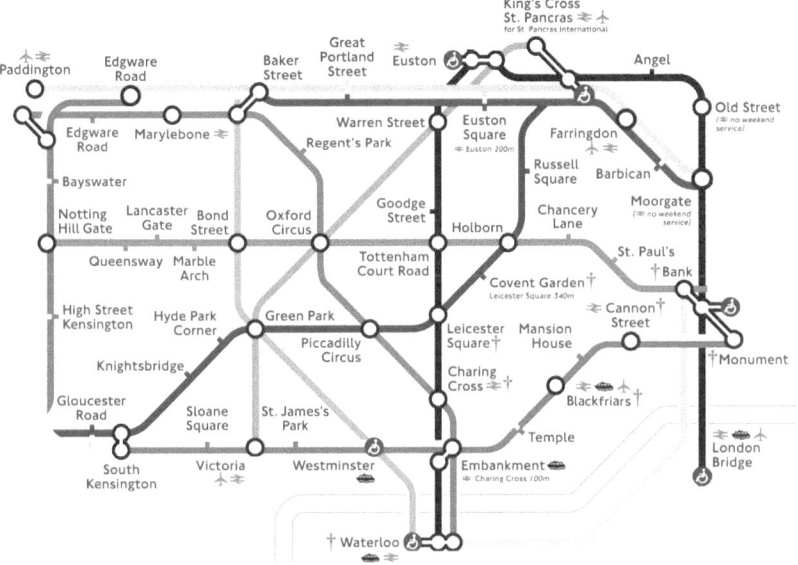

Fig. 7 Central London Tube map (*Source* Faramondi et al., 2020)

proposed methodology allows the decision maker to prioritize the protection of selected nodes, based on several criteria at once and considering their subjective preferences among the criteria (Faramondi et al., 2020).

4 Discussion and Conclusions

To conclude, let us now briefly discuss the relation between human decision-making and automatic decision-making, with particular reference to the AHP methodology. In several contexts, mere automatic decision-making is impossible. In order to take effective decisions, we need to complement data-driven approaches (e.g., machine learning) with experience-driven approaches that attempt to incorporate the actual opinion of decision makers in the automatic decision framework or expert system. Indeed, combining these approaches would allow a complete view of the phenomenon.

We observe a dichotomy: data-driven decision-making approaches are able to match inputs (e.g., the alternatives to compare) and outputs (e.g.,

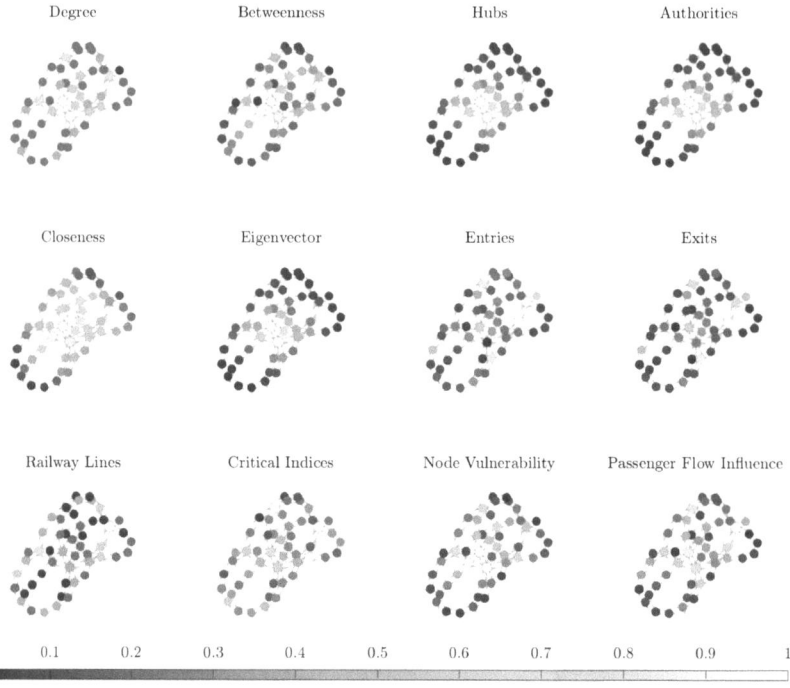

Fig. 8 Evaluation of the relevance of nodes, taking twelve metrics into account. The evaluations are normalized in the range [0, 1], where 0 and 1 indicate minimum and maximum relevance, respectively (*Source* Faramondi et al., 2020)

the choice) based on the availability of a large number of examples. Moreover, they represent quite effective ways to interpolate among known cases. Yet, such systems are, typically, black boxes that do not enable an understanding of how choices are made; in other words, no new knowledge is generated that fosters the comprehension of the underlying phenomena.

In this view, experience-driven decision-making represents a complementary perspective. Such approaches, typically, construct a model based on the actual preferences of a decision maker: as a result, we can look inside the black box. Although effective, such techniques can be implemented at the cost of handling ambiguity, contradiction and inconsistencies. However, in this case, new knowledge is obtained, in that the system replicates the experts' way of thinking.

We argue that, in order to obtain the best from both worlds, a combination of these diverse approaches is needed, especially when there is a need to solve complex problems, where there is only partial availability of objective data, for example, in expert medical systems, where we want to consider the human experience of diagnosis or critical infrastructure protection, where "black swan" events are unprecedented if you rely only on past event data without leveraging the experience of human operators and experts.

In this context, the AHP methodology represents a perfect case study highlighting how imprecise relative information provided by human experts can be transformed into actual indications on the absolute relevance or value of the alternatives.

Appendix

This appendix formally presents the key mathematical concepts discussed in the chapter. We will discuss the main machinery underlying the incomplete analytic hierarchy process problem, with particular reference to the logarithmic least-squares method. The interested reader is referred to Godsil and Royle (2001), Bozóki and Tsyganok (2019), Faramondi et al. (2020) and Oliva et al. (2019) for more details.

General Notation

Vectors are denoted in bold, while matrices are shown with upper-case letters. We use A_{ij} to address the (i, j)-th entry of a matrix A and x_i for the i-th entry of a vector $\boldsymbol{x}$. Moreover, we write 1_n and 0_n to denote a vector with n components, all equal to one and zero, respectively; similarly, we use $1_{n \times n}$ and $0_{n \times n}$ to denote $n \times m$ matrices all equal to one and zero, respectively. We denote by $\boldsymbol{I}_n$ the $n \times n$ identity matrix. Finally, we express by $\exp(\boldsymbol{x})$ and $\ln(\boldsymbol{x})$ the component-wise exponentiation or logarithm of the vector $\boldsymbol{x}$, i.e., a vector such that $\exp(\boldsymbol{x})_i = e^{x_i}$ and $\ln(\boldsymbol{x})_i = \ln(x_i)$, respectively.

Fig. 9 Node importance according to the holistic index based on LLS AHP (*Source* Faramondi et al., 2020)

GRAPH THEORY

Let $G = \{V, E\}$ be a *graph* with n nodes $V = \{v_1, \ldots, v_n\}$ and e edges

$$E \subseteq V \times V \{(v_i, v_j) | v_i \in V\},$$

where $(v_i, v_j) \in E$ captures the existence of a link from node v_i to node v_j. A graph is said to be *undirected* if $(v_i, v_j) \in E$ whenever $(v_j, v_i) \in E$, and is said to be *directed* otherwise. In the following, when dealing with undirected graphs, we represent edges using unordered pairs $\{v_i, v_j\}$ in place of the two directed edges $(v_i, v_j), (v_j, v_i)$. A graph is *connected* if for each pair of nodes v_i, v_j there is a path over G that connects them. Let the neighbourhood $\mathcal{N}_i$ of a node v_i in an undirected graph G be the set of nodes v_j that are connected to v_i via an edge $\{v_i, v_j\} \in E$. The *degree*

d_i of a node v_i in an undirected graph G is the number of its incoming edges, i.e., $d_i = |\mathcal{N}_i|$. The *degree matrix* D of an undirected graph G is the $n \times n$ diagonal matrix such that $D_{ii} = d_i$. The *adjacency matrix* Adj of a directed or undirected graph $G = \{V, E\}$ with n nodes is the $n \times n$ matrix such that $\text{Adj}_{ij} = 1$ if $\{v_i, v_j\} \in E$ and $\text{Adj}_{ij} = 0$, otherwise. The *Laplacian matrix* associated with an undirected graph G is the $n \times n$ matrix L, having the following structure

$$L_{ij} = \begin{cases} 1, & \text{if } \{v_i, v_j\} \in E, \\ d_i, & \text{if } i = j, \\ 0, & \text{otherwise.} \end{cases}$$

It is well known that L has an eigenvalue equal to zero, and that, in the case of undirected graphs, the multiplicity of such an eigenvalue corresponds to the number of connected components of G (Godsil & Royle, 2001). Therefore, the eigenvalue zero has multiplicity one if and only if the graph is connected.

INCOMPLETE ANALYTIC HIERARCHY PROCESS

In this subsection we review the analytic hierarchy process (AHP) problem when the available information is incomplete.

The aim is to compute an estimate of the unknown utilities, based on information on relative preferences. To this end, consider a set of n alternatives, and suppose that each alternative is characterized by an unknown utility or value w_i 0. In the incomplete information case, we are given a value $\mathcal{A}_{ij} = \frac{\epsilon_{ij} w_i}{w_j}$ for selected pairs of alternatives i, j; such a piece of information corresponds to an estimate of the ratio w_i/w_j, where ϵ_{ij} 0 is a multiplicative perturbation that represents the estimation error. Moreover, for all the available entries $\mathcal{A}_{ij}$, we assume that $\mathcal{A}_{ji} = \mathcal{A}_{ij}^{-1} = \frac{\epsilon_{ij}^{-1} w_j}{w_i}$, i.e., the available terms $\mathcal{A}_{ij}$ and $\mathcal{A}_{ji}$ are always consistent and satisfy $\mathcal{A}_{ij} \mathcal{A}_{ji} = 1$.

We point out that, while traditional AHP approaches (Barzilai et al., 1987; Crawford, 1987; Saaty, 1977) require knowledge of every pair of alternatives, in the partial information setting we are able to estimate the vector $\mathbf{w} = [w_1, \ldots, w_n]^T$ of the utilities, knowing just a subset of the perturbed ratios. Specifically, let us consider a graph $G = \{V, E\}$, with $|V| = n$ nodes; in this view, each alternative i is associated with a node

$v_i \in V$, while the knowledge of $w_{i\,j}$ corresponds to an edge $(v_i, v_j) \in E$. Clearly, since we assume knowing w_{ij} whenever we know w_{ji}, the graph G is undirected. Let $\mathcal{A}$ be the $n \times n$ matrix collecting the terms $\mathcal{A}_{ij}$, with $\mathcal{A}_{ij} = 0$ if $(v_i, v_j) \notin E$. Notice that, in the AHP literature, there is no universal agreement on how to estimate the utilities in the presence of perturbations (see for instance the debate in Dyer (1990), Saaty (1990), for the original AHP problem). This is true also in the incomplete information case; see, for instance, Bozóki et al. (2010), Oliva et al. (2017), Menci et al. (2018). While the debate is still open, we point out that the logarithmic least-squares approach appears particularly appealing, since it focuses on error minimization.

For these reasons, we now review the incomplete logarithmic least-squares (ILLS) method (Bozóki et al., 2010; Menci et al., 2018), which represents an extension of the classical logarithmic least-squares (LLS) method developed in Crawford (1987), Barzilai et al. (1987), for solving the AHP problem in the complete information case.

INCOMPLETE LOGARITHMIC LEAST-SQUARES APPROACH TO AHP

Within the ILLS algorithm, the aim is to find a logarithmic least-squares approximation w^* to the unknown utility vector w, i.e., to find the vector that solves

$$\mathbf{w}^* = \arg\min_{x \in \mathbb{R}^n_+} \left\{ \frac{1}{2} \sum_{i=1}^{n} \sum_{j \in \mathcal{N}_i} \left(\ln(\mathcal{A}_{ij}) - \ln\left(\frac{x_i}{x_j}\right) \right)^2 \right\}.$$

An effective strategy to solve the above optimization problem is to operate the substitution $y = \ln(x)$, where $\ln(\cdot)$ is the component-wise logarithm, so that the above equation can be rearranged as

$$\mathbf{w}^* = \exp\left(\arg\min_{y \in \mathbb{R}^n} \left\{ \frac{1}{2} \sum_{i=1}^{n} \sum_{j \in \mathcal{N}_i} \left(\ln(\mathcal{A}_{ij}) - y_i + y_j \right)^2 \right\} \right),$$

where $\exp(\cdot)$ is the component-wise exponential. Let us define

$$\kappa(\mathbf{y}) = \frac{1}{2} \sum_{i=1}^{n} \sum_{j \in \mathcal{N}_i} \left(\ln(\mathcal{A}_{ij}) - y_i + y_j \right)^2;$$

because of the substitution $y = \ln(x)$, the problem becomes convex and unconstrained, and its global minimum is in the form $w = \exp(y^*)$, where y^* satisfies

$$\left.\frac{\partial \kappa(y)}{\partial y}\right|_{y=y^*} = \sum_{j \in N_i} \left(\ln(\mathcal{A}_{ij}) - y_i + y_j\right) = 0, \forall i = 1, \ldots, n,$$

i.e., we seek the argument of the function that nullifies its derivative. Let us consider the $n \times n$ matrix P such that $P_{ij} = \ln(\mathcal{A}_{ij})$ if $\mathcal{A}_{ij} > 0$ and $P_{ij} = 0$, otherwise; we can express the above conditions in a compact form as

$$L y^* = P 1_n,$$

where L is the Laplacian matrix associated with the graph G. Note that, since for hypothesis G is undirected and connected, the Laplacian matrix L has rank $n - 1$ (Godsil & Royle, 2001). Therefore, a possible way to calculate a vector y^* that satisfies the above equation is to fix one arbitrary component of y^* and then solve a reduced size system simply by inverting the resulting nonsingular $(n - 1) \times (n - 1)$ matrix (Bozóki, & Tsyganok, 2019).

Vector y^* can also be written as the arithmetic mean of vectors calculated from the spanning trees of the graph of comparisons, corresponding to the incomplete additive pairwise comparison matrix $\ln(\mathcal{A})$(Bozóki, & Tsyganok, 2019). Finally, it is worth mentioning that, when the graph G is connected, the differential equation

$$\dot{y}(t) = -L y(t) + P 1_n$$

asymptotically converges to y^* (see Olivati-Saber et al., 2007), representing yet another way to compute it. Notably, the latter approach is typically used by the control system community for formation control of mobile robots, since the computations are easily implemented in a distributed way and can be performed cooperatively by different mobile robots. Therefore, such an approach appears particularly appealing in a distributed computing setting.

Merging Multiple Opinions

In view of the developments in this paper, we now provide a way to calculate a ranking for a group of decision makers, each with its own perturbed ratio matrix $\mathcal{A}^{(u)}$ which does not necessarily correspond to a connected graph. To do this consider m decision makers and suppose each decision maker u provides an $n \times n$ possibly perturbed sparse ratio matrix $\mathcal{A}^{(u)}$, which has the same structure as a possibly disconnected graph $G^{(u)} = \{V, E^{(u)}\}$. Denote by

$$G = \left\{ V, \bigcup_{u=1}^{m} E^{(u)} \right\}$$

the graph corresponding to the overall information provided by the m decision makers (i.e., a graph featuring the union of the edges provided by all decision makers, where repeated edges are allowed), and consider the optimization problem

$$\mathbf{w}^* = \exp\left(\underset{\mathbf{y} \in \mathbb{R}^n}{\arg\min} \left\{ \frac{1}{2} \sum_{u=1}^{m} \sum_{i=1}^{n} \sum_{j \in \mathcal{N}_i} \left(\ln\left(\mathcal{A}_{ij}^{(u)}\right) - y_i + y_j \right)^2 \right\} \right),$$

the global optimal solution to the above problem $\mathbf{y}^*$ satisfies

$$\sum_{u=1}^{m} L(G^{(u)}) \mathbf{y}^* = \sum_{u=1}^{m} P^{(u)} 1_n$$

where $L(G^{(u)})$ is the Laplacian matrix associated with $G^{(u)}$ and $P^{(u)}$ is an $n \times n$ matrix collecting the logarithm of the non-zero entries of $\mathcal{A}_{ij}^{(u)}$, while $P^{(u)} = 0$ when $\mathcal{A}_{ij}^{(u)} = 0$. Moreover $\exp(\mathbf{y}^*)$ is unique up to a scaling factor if and only if G is connected. Looking in greater detail, we observe that the problem is an unconstrained convex minimization problem; therefore, by evaluating the derivative of the objective function at zero, we find that the optimal solution $\mathbf{y}^*$ satisfies the above equation.

REFERENCES

Barzilai, J., Cook, W. D., & Golany, B. (1987). Consistent weights for judgements matrices of the relative importance of alternatives. *Operations Research Letters, 6*(3), 131–134.

Bozóki, S., & Tsyganok, V. (2019). The (logarithmic) least squares optimality of the arithmetic (geometric) mean of weight vectors calculated from all spanning trees for incomplete additive (multiplicative) pairwise comparison matrices. *International Journal of General Systems, 48*(4), 362–381.

Bozóki, S., Fülöp, J., & Rónyai, L. (2010). On optimal completion of incomplete pairwise comparison matrices. *Mathematical and Computer Modelling, 52*(1–2), 318–333.

Coombs, C. H. (1958). On the use of inconsistency of preferences in psychological measurement. *Journal of Experimental Psychology, 55*(1), 1.

Crawford, G. (1987). The geometric mean procedure for estimating the scale of a judgement matrix. *Mathematical Modelling, 9*(3–5), 327–334.

Dyer, J. S. (1990). Remarks on the analytic hierarchy process. *Management Science, 36*(3), 249–258.

Faramondi, L., Oliva, G., & Setola, R. (2020). Multi-criteria node criticality assessment framework for critical infrastructure networks. *International Journal of Critical Infrastructure Protection, 28*, 100338.

Godsil, C., & Royle, G. (2001). *Algebraic graph theory*. Graduate text in mathematics. Springer.

Golden, B., Wasil, E., & Harker, P. (Eds.). (1989). *The analytic hierarchy process*. Springer.

Hill, R. J. (1953). A note on inconsistency in paired comparison judgments. *American Sociological Review, 18*(5), 564–566.

Ho, W. (2008). Integrated analytic hierarchy process and its applications – A literature review. *European Journal of Operational Research, 186*(1), 211–228.

Hoch, S. J., & Loewenstein, G. F. (1991). Time-inconsistent preferences and consumer self-control. *Journal of Consumer Research, 17*(4), 492–507.

Menci, M., Oliva, G., Papi, M., Setola, R., & Scala, A. (2018). A suite of distributed methodologies to solve the sparse analytic hierarchy process problem. *Proceedings of the 2018 European Control Conference*, 1147–1453.

Olfati-Saber, R., Fax, J. A., & Murray, R. M. (2007). Consensus and cooperation in networked multi-agent systems. *Proceedings of the IEEE, 95*(1), 215–233.

Oliva, G., Faramondi, L., Setola, R., Tesei, M., & Zio, E. (2021). A multi-criteria model for the security assessment of large-infrastructure construction sites. *International Journal of Critical Infrastructure Protection, 35*, 100460.

Oliva, G., Scala, A., Setola, R., & Dell'Olmo, P. (2018). Data for: opinion-based optimal group formation. *Mendeley Data*. https://doi.org/10.17632/b3ds68ygt6.1

Oliva, G., Scala, A., Setola, R., & Dell'Olmo, P. (2019). Opinion-based optimal group formation. *Omega, 89*, 164–176.

Oliva, G., Setola, R., & Scala, A. (2017). Sparse and distributed analytic hierarchy process. *Automatica, 85*, 211–220.

Regenwetter, M., & Davis-Stober, C. P. (2012). Behavioral variability of choices versus structural inconsistency of preferences. *Psychological Review, 119*(2), 408.

Rodrigues, F. A. (2019). *Network centrality: An introduction*. Springer.

Saaty, T. L. (1977). A scaling method for priorities in hierarchical structures. *Journal of Mathematical Psychology, 15*(3), 234–281.

Saaty, T. L. (1990). An exposition of the AHP in reply to the paper "remarks on the analytic hierarchy process." *Management Science, 36*(3), 259–268.

Saaty, T., & Vargas, L. (2012). *Models, methods*. Springer.

Subramanian, N., & Ramanathan, R. (2012). A review of applications of analytic hierarchy process in operations management. *International Journal of Production Economics, 138*(2), 215–241.

Vaidya, O., & Kumar, S. (2006). Analytic hierarchy process: An overview of applications. *European Journal of Operational Research, 169*(1), 1–29.

CHAPTER 6

Depression Detection: Text Augmentation for Robustness to Label Noise in Self-Reports

Javed Ali, Dat Quoc Ngo, Aninda Bhattacharjee, Tannistha Maiti, Tarry Singh, and Jie Mei

Abstract With a high prevalence in both high- and low-to-middle-income countries, depression is one of the world's most common mental disorders, placing heavy burdens on society. Depression severely impairs the daily functioning and quality of life of individuals of different ages and may eventually lead to self-harm and suicide. Advancements have emerged

J. Ali · D. Q. Ngo · A. Bhattacharjee · T. Maiti (✉) · T. Singh · J. Mei
deepkapha.ai, Amsterdam, The Netherlands
e-mail: tannistha.maiti@deepkapha.com

J. Ali
e-mail: javed.ali@deepkapha.ai

D. Q. Ngo
e-mail: dat.ngo@deepkapha.com

A. Bhattacharjee
e-mail: aninda.bhattacharjee@deepkapha.com

© The Author(s), under exclusive license to Springer Nature
Switzerland AG 2022
M. Bertolaso et al. (eds.), *Digital Humanism*,
https://doi.org/10.1007/978-3-030-97054-3_6

81

in the fields of natural language for depression assessment, using bidirectional encoder representation from transformers (BERT). In this study, we used the Reddit Self-reported Depression Diagnosis dataset, an uncurated text-based dataset, to enable the detection of depression using easily accessible data and without the intervention of domain experts. To reduce the negative impact of label noise on the performance of transformers-based classification, we proposed two data augmentation approaches, Negative Embedding and Empathy for BERT and DistilBERT, to exploit the usage of pronouns and affective, depression-related words in the dataset. The use of Negative Embedding improves the accuracy of the model by 31% compared with a baseline BERT and a DistilBERT, whereas Empathy underperforms baseline methods by 21%. Taken together, we argue that the detection of depression can be performed with high accuracy on datasets with label noise using various augmentation approaches and BERT.

Keywords Depression · Reddit Self-reported Depression Diagnosis dataset · Clustering · Empathy · Negative Embeddings

1   INTRODUCTION

Depression is a mental health disorder that affects more than 264 million people worldwide (James et al., 2018). Patients with depression show emotional and cognitive alterations (Haro, 2019; Rao et al., 1991), with a wide range of symptoms varying from trouble concentrating, remembering details (Kreutzer et al., 2001) and making decisions to feelings

T. Singh
e-mail: tarry.singh@deepkapha.com

J. Mei
e-mail: jie.mei@deepkapha.com

J. Ali · D. Q. Ngo
Indian Institute of Technology, Kharagpur, India

J. Mei
University of Western Ontario, London, Canada

of guilt, worthlessness and helplessness. Depression can lead to suicidal thoughts and attempts (Pedersen, 2008; Toolan, 1962): according to the WHO, approximately 800,000 people die due to suicide every year, and suicide is the second leading causes of death in adolescents and young adults. Common mental disorders like depressive and anxiety disorders are frequently encountered in primary care (Avasthi & Ghosh, 2014). However, recognition of these disorders is poor, with less than a third of clinically significant cases being identified (Avasthi & Ghosh, 2014; Patel & Bakken, 2010). Currently, the COVID-19 pandemic is leading to psychological distress in the general public, and therefore, a public mental health crisis (Campion et al., 2020; Lu & Bouey, 2020; Pfefferbaum & North, 2020). Studies indicate that people who did not have sufficient daily supplies during the lockdown were most affected, and family wealth was found to be negatively correlated with stress, anxiety and depression (Dutta & Bandyopadhyay 2020; Rehman et al., 2020). In the meantime, given that access to mental healthcare facilities is highly limited (Moreno et al., 2020), early detection of depression is of great importance during the pandemic to provide timely treatment and to prevent suicide and attempted suicide (Losada et al., 2020).

Various studies have applied machine-learning analysis to the detection of the early signs of depression. Losada et al. (2017) and Cacheda et al. (2018) explored issues of evaluation methodology, effectiveness metrics and other processes related to early risk detection. Much of the work was focused around lexical, linguistic, semantic, or statistical analysis, textual similarity and writing features. Lam et al. (2019) proposed machine-learning approaches on the Distress Analysis Interview Corpus through feature engineering for acoustic feature modelling. Zulfiker et al. (2021) proposed a study to investigate different machine-learning classifiers using various socio-demographic and psychosocial information to detect whether a person is depressed or not. They used features like minimum redundancy, and maximum and Boruta features to extract the most relevant features from the dataset. Other research developed useful diagnostic biomarkers from brain imaging data to detect major depressive disorder (MDD) in its early stages. Mousavian et al. (2021) develop a connectivity- based machine-learning (ML) workflow that utilizes similarity/dissimilarity of spatial cubes in brain MRI images as features for depression detection.

With rapidly increasing internet use, people now have more opportunities to share their stories, challenges and mental health problems through online platforms such as Reddit and Twitter (Burdisso et al., 2019; Naslund et al., 2020). Accordingly, analysis of text data provides a new avenue for the understanding and early detection of depression. Using natural language processing (NLP) techniques and machine-learning algorithms, researchers have proposed novel approaches to the diagnosis of mental disorders including depression (Benton et al., 2017; Coppersmith et al. 2015a; Maupomé & Meurs, 2018; Nadeem, 2016; Paul et al., 2018; Resnik et al., 2015).

In recent years, bidirectional encoder representations from transformers (BERT) (Devlin et al., 2018) has become a widely used language model extensively implemented by researchers to achieve state-of-the-art performance in various language understanding tasks. As BERT is composed of attention-based transformer blocks and pre-trained on large corpora, it can capture a variety of linguistic features and contexts and can be fine-tuned for downstream tasks for higher performance. Given the proven performance of BERT, we used it as our baseline method in the classification of depressive and non-depressive statements. Web-scraping with fixed labelling rules is a common approach for building large-scale text datasets for the diagnosis of depression (AlSagri & Ykhlef, 2020; Shen et al., 2017). Similarly, we built our Reddit Self-reported Depression Diagnosis (RSDD) dataset by web-scraping depressive and non-depressive statements from two subreddits, */depression* and */AskReddits*. A drawback of this method is label noise, that is, mislabeling by non-experts or oversimplified labelling criteria. Oversimplified labeling criteria may lead to mislabeling of a non-depressive statement in the /depression subreddit as depressed. Due to pattern-memorization effects, label noise may significantly compromise the performance of deep-learning models in classification tasks (Flatow & Penner, 2017; Zhu & Wu, 2004), particularly in the detection of depression.

The present study performs two tasks: (a) unsupervised learning to remove noise from the dataset and (b) exploitation of contexts for robustness to label noise in the detection of depression using two data augmentation methods, Negative Embedding and Empathy. We used the RSDD dataset to evaluate and demonstrate the performance of the two proposed methods in improving diagnostic accuracy for the diagnosis of depression and experimented with the two augmentation methods with both BERT and DistilBERT (Sanh et al., 2019) to understand the impacts of model distillation on the performance of the two augmentation methods.

2 Background

Table 1 provides analysis of previously published depression datasets, which will be further elaborated in the following sections.

The dataset from Computational Linguistics and Clinical Psychology (CLPsych) 2015 (Coppersmith et al., 2015a) was compiled from Twitter users who stated a diagnosis of depression or post-traumatic stress disorder (PTSD) and includes demographically matched community controls. Coppersmith et al. (2015b) also note that many users tweet statements of diagnosis, such as "I was just diagnosed with X and...", where X is a mental health condition. Depression or PTSD are among the most common found on Twitter and have a relatively high prevalence compared to other conditions. One of the limitations of the CLPsych dataset is that it needs a human annotator to evaluate each such statement of diagnosis to remove jokes, quotes or any other disingenuous statements.

The Self-reported Depression Diagnosis (RSDD) dataset (Yates et al., 2017) from Reddit is also a widely used dataset that is generated from publicly available Reddit forums. The most common false positives included hypotheticals (e.g., "if I was diagnosed with depression"), negations (e.g., "it's not like I've been diagnosed with depression"), and quotes (e.g., "my brother announced 'I was just diagnosed with depression'"). The diagnosed users are matched with control users who are interested in similar subreddits and have similar activity levels, preventing biases based on the subreddits users are involved in or based on how active the users are on Reddit. The posts were annotated to confirm that they contained claims of a diagnosis as well as control users that were matched with each diagnosed user.

CLEF16 is another publicly available dataset on depression and language uses (Losada & Crestani, 2016) consisting of a series of textual interactions written by different subjects. Losada and Crestani (2016) processed the dataset based on (i) the size and quality of the data sources, (ii) the availability of a sufficiently long history of interactions of the individuals in the collection and iii) the difficulty of distinguishing depressed cases from non-depressed cases. One of the advantages of this dataset is that it consists of substantive content about different medical conditions, such as anorexia or depression.

Table 1 Systematic analysis of published depression datasets from social media

Dataset and Year	Data source	Language	Limitation	Size	Application
CLPsych, 2015	Twitter	English	Human annotator evaluates each such statement of diagnosis to remove jokes, quotes or any other disingenuous statements	Total 1,146 users; for each user, up to their most recent 3000 public tweets included	Mental health severity detection
Self-reported Depression Diagnosis (RSDD), 2017	Reddit	English	The posts were annotated to confirm that they contained claims of a diagnosis as well as control users that were matched with each diagnosed user	9,000 diagnosed users'and 107,000 matched control users' posts	Scalable mental health analysis
CLEF16, 2016	Twitter and Reddit	English	Human annotator evaluator is required	531,453 submissions from 892 unique users	To check people with suicidal inclinations, or people susceptible to depression
Dataset of depressive and suicidal posts, 2019	Vkontakte social network	Russian	The posts were searched for keywords and then annotated using LSTM and BiLSTM	32,018 depressive postsand 32,021 usual posts	To calculate risk factors for suicide
Dataset for Research on Depression in Social Media, 2020	Reddit	English	Different heuristics were used to filter the dataset	3,500D + postsand 3,500D- posts	To understand when sadness ordisgust emotions are present

A dataset of depressive and suicidal posts (Narynov et al., 2020) was collected from Vkontakte social network using VK.api. A Python framework was used to collect the dataset and then filtering was done by applying the most used keywords that would signify depressive mood. One of the advantages of this dataset is that it also contains the age of the post author which can help improve machine-learning models.

A Dataset for Research on Depression in Social Media by Ríssola et al. (2020), proposes a method to automatically generate the large number of datasets of depression and non-depression posts from social media. This dataset is only available on request. The data were taken from the eRisk2018 collection which contains posts from two groups of Reddit users, depressed and non-depressed. The posts were assigned two scores, a Sentiment Polarity Score and a Topical Polarity Score. The advantage of this dataset is that the authors have provided a benchmark to compare future works.

All the datasets discussed above are potential approaches to learning with labels. Traditionally, data cleaning has been applied, which relied on finding heuristic points that were corrupted by label noise and filtering them out (Angelova et al., 2005; Brodley & Friedl, 1999). Current techniques focus on improving learning algorithms and modifying neural network architectures for estimating the true labels based on noisy labels, for example, using bootstrapping to combine multiple weak models trained on k folds of data into a strong model to learn under label noise (Algan & Ulusoy, 2020).

As stated above, large datasets in NLP suffer from noisy labels, due to erroneous automatic and human annotation procedures. For language-based tasks, recent studies addressed the issue of label noise by supplying additional contextual information to the attention models. The attention mechanism individually computes attention weights of each token over the bag-of-word tokens (Vaswani et al., 2017). As a result, attention models such as generative pre-trained transformer (GPT) (Radford et al., 2019) and BERT (Devlin et al., 2018) neglect contextual information in the calculation of dependencies between tokens (Yang et al., 2019).

Table 2 Samples of the Reddit Self-reported Depression Diagnosis (RSDD) dataset for/depression and/Askreddit. The first comment in/depression is a non-depressive sentence. This is an example of label noise

/Depression		/AskReddit	
Title	*Comments*	*Title*	*Comments*
I am so tired of people taking me for granted I give them too much of energy. I am sick of everything. My life, my family, my friends	– I'm sorry. I'm really hoping the best for you – I know how you feel. I feel exactly the same right now. I wish I could give this post a thousand rewards	What's something that impresses most people that doesn't impress you?	– Limousines. As a kid, I used to think that was the sign that you made it. Now I realize you just need $95 – If you've got more than 5 people, getting a limo or party bus is miles cheaper than getting multiple Ubers. Plus you can drink in them

3   Reddit Self-Reported Depression Diagnosis (RSDD) Dataset

Our present dataset utilized the Python Reddit API Wrapper to web-scrape posts from January 2018 to November 2020 in the two subreddits */depression* and */AskReddit*, which correspond to "depressed" and "non-depressed" classes, respectively. The title and comments of each post were then anonymized and treated as separate samples. The total number of text samples[1] was 229,729 (depressed:102,102; non-depressed: 127,627), and for each class, text samples were split by a ratio of 80:20 for training and validation sets. Table 2 shows an example from the "depressed" category dataset and its comments. A challenge in this web-scraped dataset is label noise: positive and supportive statements (e.g.,"I'm sorry. I'm really hoping the best for you") in the

---

[1] Dataset is included for submission {https://drive.google.com/drive/folders/1_Hob3 6rauDlrvvgu7Umxs6HV9sqHzlUB?usp=sharing}

Fig. 1 Word cloud demonstrating the frequency distribution of words in (a) depressive samples of the RSDD dataset, (b) non-depressive samples of the RSDD data

/depression subreddit are labelled as "depressed" during the automatic web-scraping process. Without proper annotation, the correctness of the labels is compromised, and the dataset is prone to noise. Such data is defined as labelled noise. The performance of machine-learning models hugely depends on the data and the corresponding label.

Words such as "feel", "depression", "want" and "friend" are seen more frequently in the depressive samples (see word cloud in Fig. 1). The word cloud of the lexicons generated by the Empath library, as shown in Fig. 2, demonstrates that the depressive samples include more affective words. Pronouns are also commonly observed in depressive samples.

The dataset for this study will be freely released for other researchers to use. However, the data will in future be protected by some form of data compliance EULA (end-user licence agreement) that addresses relevant compliance and the risk of the dataset being used other than for research.

Preprocessing and Noise Removal

Word embedding is a term used for the representation of words for text analysis, typically in the form of a real-valued vector that encodes the meaning of the word such that the words that are closer in the vector space are expected to be similar in meaning. After generating the datasets,

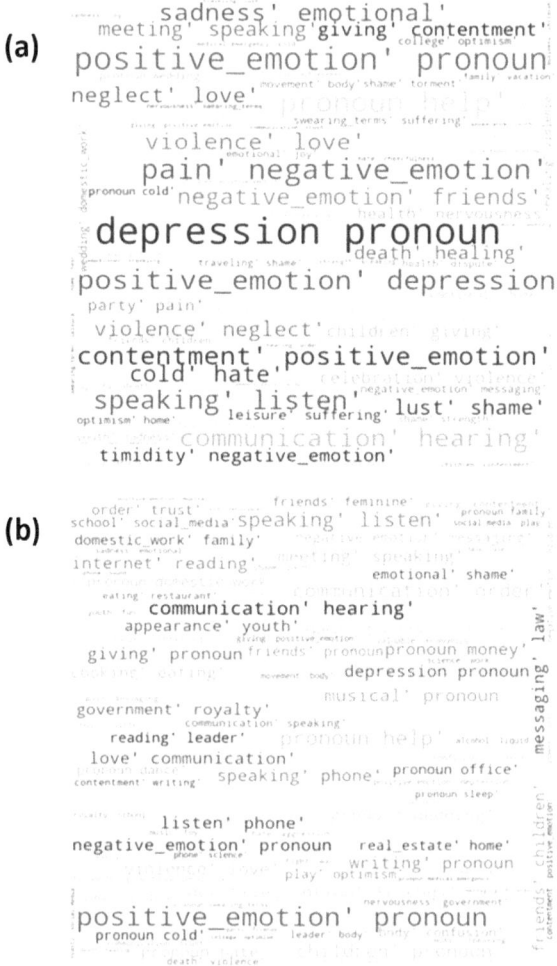

Fig. 2 Word cloud demonstrating the frequency distribution of words in (a) lexicons generated by the Empath library from the depressive samples of the RSDD dataset, (b) lexicons generated by the Empath library from the non-depressive samples of the RSDD dataset

we converted our raw data into numerical representation using the Google Universal Sentence Encoder (GUSE) word-embedding model. The Universal Sentence Encoder encodes text into high-dimensional vectors that can be used for text classification, semantic similarity, clustering, and other natural language tasks. The GUSE model is trained and optimized for longer texts like phrases, sentences and short paragraphs.

To remove the noise from our dataset, we used various unsupervised clustering algorithms. Since the data, after GUSE embedding, was very high-dimensional, we first applied dimensionality reduction algorithms. Dimensionality reduction algorithms are necessary to avoid the poor performance of clustering due to dimensionality. Principal component analysis (PCA) is a technique used for reducing the dimensionality of datasets, increasing interpretability but at the same time minimizing information loss. It does so by creating new uncorrelated variables that successively maximize the variance. Through this, we reduced the dimension of word embedding to two.

After completing dimensionality reduction, we applied various clustering algorithms: Gaussian mixture models (GMM) and K-means clustering. GMM is a probabilistic model that assumes all the data points are generated from a mixture of a finite number of Gaussian distributions with unknown parameters. K-means is a centroid-based algorithm, or a distance-based algorithm, where we calculate the distances to assign a point to a cluster. In K-means, each cluster is associated with a centroid. Through these clustering algorithms, we divided our data into two clusters. These two clusters were assigned as labels to our model.

After the clustering, we had two types of labels for each post, one based on which subreddit it was scraped from and the other based on the clustering algorithm. To decide which one to use, we used thresholding-based label correction. For this method we checked the confidence of the clustering algorithm: if the probability of a label was greater than 90%, then we used the clustering algorithms label, otherwise we used the original label (Fig. 3).

Empathy

An alternative approach is to generate high-level lexicons which represent the overall emotional context.

Researchers have relied on such high-level lexicons to identify signs of depression in social media posts and to understand the overall meaning

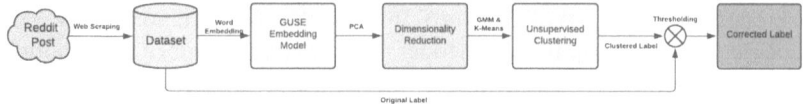

Fig. 3 Pipeline for label correction. It starts with extraction of the dataset followed by GUSE embedding model, applying dimensionality reduction, unsupervised clustering and finally label correction

of texts at scale. One of the most commonly used libraries is Linguistic Inquiry and Word Count (LIWC) which counts words relevant to lexical categories such as sadness, health, and positive emotions (Tausczik & Pennebaker, 2010). For example, positive lexicons include words such as "happy", "joy", "fun", etc. In published work (Shen et al., 2017), LIWC was used to generate lexicons as high-level text features for logistic regression models to classify depression in social media posts. LIWC has a fixed list of 40 lexical categories that limits its ability to capture signs of depression in text data.

Unlike LIWC, the Empath library is designed using deep-learning techniques and crowdsourcing that allow it to incorporate new lexical categories (Fast et al., 2016). In the present study, the proposed data augmentation method Empathy utilizes the Empath library and initially updates the library with two lexicons, "pronoun" and "depression", which consider relevant words as possible indicators of depression. This process is theoretically aligned with previous findings that depressed patients use first-person singular pronouns and depression-related words more frequently than healthy controls (Rude et al., 2004). Each text sample is evaluated by the Empath library to generate high-level lexicons which are then linearly concatenated with the text sample into a new text sample (Table 3). The generated text sample consists of both original contexts and high-level, extracted emotional contexts.

4   EXPERIMENTS

In this study, BERT and DistilBERT are used as backbones for the classification module. Both architectures are multi-layer bidirectional transformer encoders that use multi-head attention. Transformer architectures (linear layer and layer normalization) are highly optimized in modern linear algebra frameworks (Sanh et al., 2019). Variations of the

Table 3 Example of empathy-generating lexicons and concatenating generated lexicons with original text

Original Text	Lexicons	Post-processed Texts
Wow. I understand that the rules are the rules, you just painted "everyone" who offers that as either a psycho or a predator. I must say I am feeling like one now because ...	hate, nervous, suffering, art, optimism, fear, zest, speaking, sympathy, sadness, joy, lust, shame, pain, negative emotion, contentment, positive emotion, depression, pronoun, ...	Wow. I understand that the rules are the rules, you just painted"everyone" who offers that as ... hate, nervous, suffering, art, optimism, fear, zest, speaking, sympathy, sadness, joy, lust, shame, pain, ...

Table 4 Model parameters used in this study for BERT and DistilBERT

Components	BERT	DistilBERT
Transform Block (L)	12	6
Hidden Size (H)	768	768
Self-attention heads (A)	12	12
Max Sequence Length	256	256

last dimension of the tensor (i.e., hidden size dimension) have less of an impact on computation efficiency for a fixed parameters budget than variations of other factors such as the number of layers. The backbones of BERT and DistilBERT are the pre-trained BERT-Base-Uncased and DistilBERT-Base-Uncased, the parameters of which are adapted from Devlin et al. (2018) and Sanh et al. (2019) (Table 4). The classification module is composed of a global averaging layer with a pool size set to 3 and a stride of 3, then two hidden fully connected layers of 256 and 64 units, each followed by a rectified linear unit (ReLU) activation function.

Negative Embedding

In the conditional masked language model (MLM) pre-training task for BERT (Wu et al., 2019), the segmentation embedding was replaced by label embedding to control word predictions on conditions of labels while preserving context. Inspired by this work, we replaced segmentation embeddings with negative embeddings in order to emphasize depressive contexts on conditions of the existence of negative tokens and to esti-mate true labels from noisy labels. The Negative Embedding labels were binary classes (1 and 0) for negative and non-negative tokens respec-tively (Fig. 2). The objective is to compute the probability of depression $p(\cdot|S\backslash\Sigma n_i)$ given the negative token n_i, the sequence S and the context $S\backslash\Sigma n_i$. The negative tokens are common negative tokens in the senti-ment analysis task and have been predefined in previous studies (Hu & Liu, 2004; Liu et al., 2005).

Hyper-Parameter Setting and Fine-Tuning

For all experiments, models were optimized with a binary cross-entropy loss function using the Adam optimizer (Kingma & Ba, 2014) with a learning rate of 0.0001 and trained with a batch size of 128. The maximum token length was set to 256. Early stopping was used to stop training after 10 epochs of no improvement in accuracy, and learning-rate scheduling was implemented to reduce the learning rate by a factor of 0.1 after 10 epochs of no reduction in loss. To leverage the pre-trained weights of BERT and DistilBERT, we initially fine-tuned the models for 5 epochs without updating backbones. Then, we fine-tuned the models for the next 50 epochs by updating backbones. All experiments were performed on an AMD Radeon VII 16 Gb GPU.

Text Data Augmentation

We tested BERT and DistilBERT with the two proposed text data augmentation techniques. For Negative Embedding, we updated BERT and DistilBERT's vocabulary with the predefined negative words in order to avoid unknown padding (Devlin et al., 2018; Sanh et al., 2019). We applied word-piece tokenization (Wu et al., 2016) to tokenize text samples and evaluated the newly formed tokens with the predefined nega-tive words to generate binary-valued negative embeddings (Fig. 2). For Empathy, we first added two depression-related lexicons ("pronoun" and "depression") to the Empath library and applied the library to analyze text and generate emotional lexicons. The lexicons were then concate-nated with original texts into new text samples (Table 3) which were

Table 5 Training and validation, evaluation metrics for two text augmentation methods, and no-augmentation on BERT and DistilBERT architectures. Val: validation set, ne: Negative Embedding

Model	Train/Val Loss	Train/Val Precision	Train/Val Recall	Train/Val Accuracy
BERT	0.54/0.54	0.77/0.77	1.0/1.0	0.77/0.77
BERT+NE	0.27/0.35	0.92/0.89	0.88/0.85	0.89/0.86
BERT+Empathy	0.69/0.69	0.56/0.56	1.00/1.00	0.56/0.55
DistilBERT	0.54/0.54	0.77/0.77	1.0/1.0	0.77/0.77
DistilBERT + NE	0.25/0.36	0.93/0.90	0.89/0.85	0.90/0.86
DistilBERT + Empathy	0.69/0.69	0.56/0.56	1.00/1.00	0.56/0.55

finally tokenized by word-piece tokenization (Wu et al., 2016). The codes used in this study have been made publicly available.[2]

5 RESULTS AND ANALYSIS

We have examined the performance of BERT and DistilBERT with or without the use of text data augmentation methods. Results demonstrate that Negative Embedding leads to great improvements in model performance and therefore, outperforms Empathy and baseline BERT and DistilBERT models in discriminating between depressive and non-depressive statements in a dataset with label noise (Table 5). The low precision and high recall of baseline models and Empathy suggest that label noise leads to more similarity in textual context among classes and causes the true non-depressive statements to be classified as depressive. While Negative Embedding achieved the highest precision, it led to the lower recall. This shows that Negative Embedding increases the contextual differences by emphasizing negative words used in depressive statements. As a result, fewer true non-depressive statements were misclassified as depressive. Another expected result of Negative Embedding was the lower recall that some non-depressive statements in the depressive group might be classified as non-depressive.

---

[2] GitHub repository: https://github.com/deepkapha/depressio.

For both BERT and DistilBERT, the use of Empathy led to low performance even compared with baseline models (Table 5). The concatenation of original texts and lexicons generated by the Empath library attempts to add high-level contextual information to the original context. However, the Empath library may generate lexicons that contradict the original context which leads to greater ambiguity in the concatenated text. In Table 3, lexicons "optimism, joy, positive emotion" are generated for texts that are overall negative. Apart from that, the depressive and non-depressive statements processed by Empathy may share many similar lexicons generated by the Empath library due to its fixed list of lexicons. This could worsen the high contextual similarity between depressive and non-depressive classes caused by label noise.

According to Table 5, the performances of BERT and DistilBERT with Negative Embedding are comparable. Figures 4 and 5 show that BERT and DistilBERT with Negative Embedding converge comparably, and converge faster than BERT and DistilBERT without text augmentation and BERT and DistilBERT with Empathy. These observations suggest that model distillation does not compromise the performance of models with Negative Embedding. During fine-tuning, loss and accuracy of BERT and DistilBERT without text augmentation and BERT and DistilBERT with Empathy do not improve (Figs. 4 and 5).

Based on Figs. 4 and 5, BERT and DistilBERT without text augmentation do not converge for a given number of epochs, and one possible reason is the generalization gap between training and test data. The model training could have benefited from a larger batch size of 4096 and the use of different learning rate schemas (Krizhevsky et al., 2012). To address this issue, the layer-wise adaptive moments optimizer (LAMB) has been proposed (You et al., 2019) which can scale the batch size of BERT pre-training to 64 K without losing accuracy. In this study, You et al. (2019) used different sequence lengths of 128 and 512. In our future studies, we will include validation of the LAMB optimizer.

The use of Negative Embedding and Empathy in BERT-based architectures on label noise is a novel approach. Earlier studies (Rodrigues Makiuchi et al., 2019) have used textual embeddings to extract BERT textual features and employ a convolutional neural network (CNN) followed by a LSTM layer. Rodrigues Makiuchi et al. (2019) trained their model on a relatively clean, labelled dataset of the Patient Health Questionnaire (PHQ). They achieved a concordance correlation coefficient (CCC) score of 0.497. Our models on negative embeddings achieved a

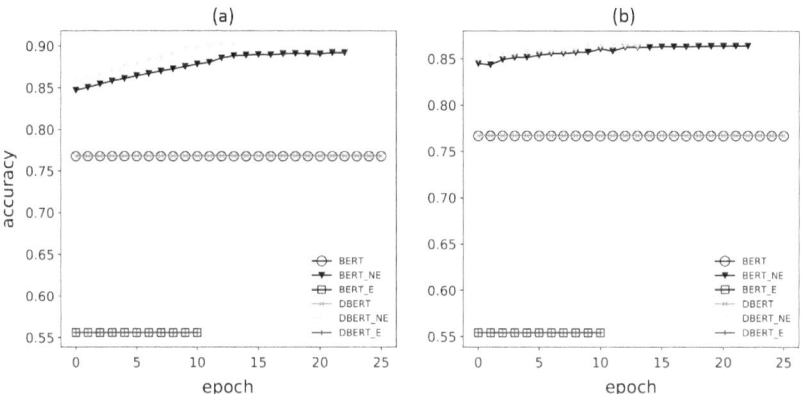

Fig. 4 Accuracy during training when backbones are unfrozen. BERT with Negative Embedding and DistilBERT with Negative Embedding have the highest accuracy on (A) training and (B) validation sets. NE: Negative Embedding, E: Empathy, DBERT: DistilBERT

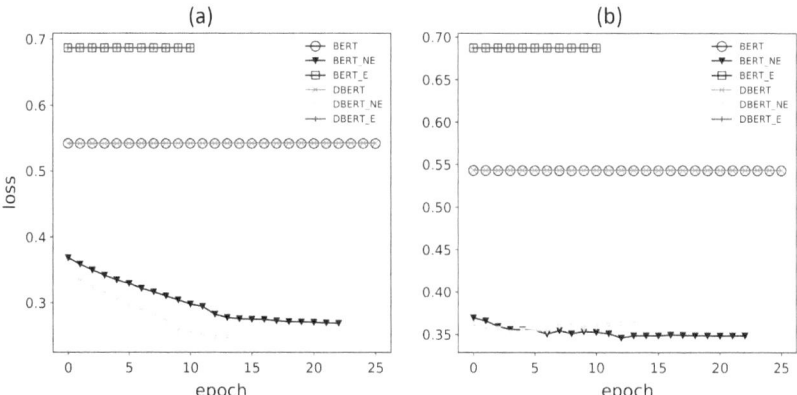

Fig. 5 Loss during training when backbones are unfrozen. BERT with Negative Embedding and DistilBERT with Negative Embedding have the lowest loss on (A) training and (B) validation sets. NE: Negative Embedding, E: Empathy, DBERT: DistilBERT

higher F1 score (approximately 87%) compared to those of Dinkel et al. (2019), where Word2Vec and fastText embeddings were used on a sparse dataset and a F1 score of 35% on average was observed.

Future studies will investigate multimodal datasets that include speech and text data which are essential in emotion recognition for the detection of depression (Siriwardhana et al., 2020). Fusion architectures with BERT have been used by Siriwardhana et al. (2020) to perform emotion recognition using speech data and to improve our study further, we will include speech data for incorporating more contextual information (Baevski et al., 2019).

Apart from the above, we will also explore knowledge-enabled bidirectional encoder representation from transformers (K-BERT) (Liu et al., 2020). K-BERT is capable of loading any pre-trained BERT models as they are identical in parameters. In addition, K-BERT can easily inject domain knowledge into the models by using a knowledge graph without pre-training. The idea behind this approach is that the detection and assessment of depression is a very domain-specific task; using a knowledge graph to inject contextual information into the BERT model may significantly improve model performance.

6   Conclusion

In this study, we investigated the negative impact of label noise on sentiment analysis for the detection of depression and investigated whether text data augmentation methods can exploit contexts for robustness to label noise. For this purpose, we created and introduced the RSDD dataset and proposed two text augmentation techniques, Negative Embedding and Empathy. Our experimental results demonstrate that Negative Embedding leads to improved performance when compared with baseline BERT and Distil-BERT models, however, the use of Empathy with these models can cause a decrease in detection accuracy. Taken together, when used with BERT and DistilBERT models, Negative Embedding exploits contextual information and improves ability to distinguish between non-depressive and depressive classes, leading to high accuracy in the detection of depression based on text data.

REFERENCES

Alambo, A., Gaur, M., & Thirunarayan, K. (2020). Depressive, drug abusive, or informative: Knowledge-aware study of news exposure during COVID-19 outbreak. *arXiv preprint* arXiv:2007.15209.

Algan, G., & Ulusoy, I. (2020). Label noise types and their effects on deep learning. *arXiv preprint* arXiv:2003.10471.

AlSagri, H. S., & Ykhlef, M. (2020). Machine learning-based approach for depression detection in twitter using content and activity features. *IEICE Transactions on Information and Systems, 103*(8), 1825–1832.

Angelova, A., Abu-Mostafam, Y., & Perona, P. (2005, June). Pruning training sets for learning of object categories. In *2005 IEEE Computer Society Conference on Computer Vision and Pattern Recognition (CVPR'05)* (Vol. 1, pp. 494–501). IEEE.

Avasthi, A., & Ghosh, A. (2014). Depression in primary care: Challenges & controversies. *The Indian Journal of Medical Research, 139*(2), 188.

Baevski, A., Auli, M., & Mohamed, A. (2019). Effectiveness of self-supervised pre-training for speech recognition. *arXiv preprint* arXiv:1911.03912.

Benton, A., Mitchell, M., & Hovy, D. (2017). Multi-task learning for mental health using social media text. *arXiv preprint* arXiv:1712.03538.

Brodley, C. E., & Friedl, M. A. (1999). Identifying mislabeled training data. *Journal of Artificial Intelligence Research, 11*, 131–167.

Burdisso, S. G., Errecalde, M., Montes-y-Gómez, M., & (2019, September). UNSL at eRisk,. (2019). *a Unified Approach for Anorexia*. Self-harm and Depression Detection in Social Media.

Cacheda, F., Iglesias, D. F., Nóvoa, F. J., & Carneiro, V. (2018). Analysis and Experiments on Early Detection of Depression. *CLEF (Working Notes), 2125*.

Campion, J., Javed, A., Sartorius, N., & Marmot, M. (2020). Addressing the public mental health challenge of COVID-19. *The Lancet Psychiatry, 7*(8), 657–659.

Coppersmith, G., Dredze, M., Harman, C., Hollingshead, K., & Mitchell, M. (2015a). CLPsych 2015 shared task: Depression and PTSD on Twitter. In *Proceedings of the 2nd Workshop on Computational Linguistics and Clinical Psychology: From Linguistic Signal to Clinical Reality* (pp. 31–39).

Coppersmith, G., Dredze, M., Harman, C., & Hollingshead, K. (2015b). From ADHD to SAD: Analyzing the language of mental health on Twitter through self-reported diagnoses. In *Proceedings of the 2nd Workshop on Computational Linguistics and Clinical Psychology: From Linguistic Signal to Clinical Reality* (pp. 1–10).

Dai, Z., Yang, Z., Yang, Y., Carbonell, J., Le, Q. V., & Salakhutdinov, R. (2019). Transformer-xl:iAttentive language models beyond a fixed-length context. *arXiv preprint* arXiv:1901.02860.

Devlin, J., Chang, M. W., Lee, K., & Toutanova, K. (2018). Bert: Pre-training of deep bidirectional transformers for language understanding. *arXiv preprint* arXiv:1810.04805.

Dinkel, H., Wu, M., & Yu, K. (2019). Text-based depression detection on sparse data. *arXiv preprint* arXiv:1904.05154.

Dong, L., & Bouey, J. (2020). Public mental health crisis during COVID-19 pandemic. *China. Emerging Infectious Diseases, 26*(7), 1616.

Dutta, S., & Bandyopadhyay, S. K. (2020). Analysis of stress, anxiety and depression of children during covid-19. *International Journal of Engineering and Management Research (IJEMR), 10*(4), 126–130.

Fana, M., Pérez, S. T., & Fernández-Macías, E. (2020). Employment impact of Covid-19 crisis: From short term effects to long terms prospects. *Journal of Industrial and Business Economics, 47*(3), 391–410.

Fast, E., Chen, B., & Bernstein, M. S. (2016, May). Empath: Understanding topic signals in large-scale text. In *Proceedings of the 2016 CHI conference on human factors in computing systems* (pp. 4647–4657).

Flatow, D., & Penner, D. (2017). On the robustness of convnets to training on noisy labels.

Haro, J. M., Hammer-Helmich, L., Saragoussi, D., Ettrup, A., & Larsen, K. G. (2019). Patient-reported depression severity and cognitive symptoms as determinants of functioning in patients with major depressive disorder: A secondary analysis of the 2-year prospective PERFORM study. *Neuropsychiatric Disease and Treatment, 15,* 2313.

Hu, M., & Liu, B. (2004). Mining and summarizing customer reviews. In *Proceedings of the tenth ACM SIGKDD international conference on knowledge discovery and data mining* (pp. 168–177).

James, O. L. I. V. E. R. (2018). Childhood relational adversity and maltreatment as the primary causes of mental disorder and distress rather than 'endogenous' genetic or neurobiological factors. *International Journal of CAT and Relational Mental Health, 2,* 9–36.

Kingma, D. P., & Ba, J. (2014). Adam: A method for stochastic optimization. *arXiv preprint* arXiv:1412.6980.

Kreutzer, J. S., Seel, R. T., & Gourley, E. (2001). The prevalence and symptom rates of depression after traumatic brain injury: A comprehensive examination. *Brain Injury, 15*(7), 563–576.

Krizhevsky, A., Sutskever, I., & Hinton, G. E. (2012). Imagenet classification with deep convolutional neural networks. *Advances in Neural Information Processing Systems, 25,* 1097–1105.

Lam, G., Dongyan, H., & Lin, W. (2019, May). Context-aware deep learning for multi-modal depression detection. In *ICASSP 2019–2019 IEEE International Conference on Acoustics, Speech and Signal Processing (ICASSP)* (pp. 3946–3950). IEEE.

Liu, B., Hu, M., & Cheng, J. (2005, May). Opinion observer: analyzing and comparing opinions on the web. In *Proceedings of the 14th international conference on World Wide Web* (pp. 342–351).

Liu, W., Zhou, P., Zhao, Z., Wang, Z., Ju, Q., Deng, H., & Wang, P. (2020, April). K-bert: Enabling language representation with knowledge graph. In *Proceedings of the AAAI Conference on Artificial Intelligence* (Vol. 34, No. 03, pp. 2901–2908).

Losada, D. E., & Crestani, F. (2016, September). A test collection for research on depression and language use. In *International Conference of the Cross-Language Evaluation Forum for European Languages* (pp. 28–39). Springer, Cham.

Losada, D. E., Crestani, F., & Parapar, J. (2017, September). eRISK 2017: CLEF lab on early risk prediction on the internet: Experimental foundations. *In International Conference of the Cross-Language Evaluation Forum for European Languages* (pp. 346–360). Springer, Cham.

Losada, D. E., Crestani, F., & Parapar, J. (2020, April). eRisk 2020: Self-harm and depression challenges. In European Conference on Information Retrieval (pp. 557-563). Springer, Cham.

Maupomé, D., & Meurs, M. J. (2018). Using Topic Extraction on Social Media Content for the Early Detection of Depression. *CLEF (Working Notes), 2125.*

Moin Nadeem. 2016. Identifying depression on Twitter. *arXiv preprint* arXiv: 1607.07384.

Moreno, C., Wykes, T., Galderisi, S., Nordentoft, M., Crossley, N., Jones, N., & Arango, C. (2020). How mental health care should change as a consequence of the COVID-19 pandemic. *The Lancet Psychiatry.*

Mousavian, M. S. (2021). Machine Learning Methods for Depression Detection Using SMRI and RS-FMRI Images.

Mousavian, M., Chen, J., Traylor, Z., & Greening, S. (2021). Depression detection from smRI and rs-fMRI images using machine learning. *Journal of Intelligent Information Systems,* 1–24.

Narynov, S., Mukhtarkhanuly, D., & Omarov, B. (2020). Dataset of depressive posts in Russian language collected from social media. *Data in Brief, 29,* 105195.

Naslund, J. A., Bondre, A., Torous, J., & Aschbrenner, K. A. (2020). Social media and mental health: Benefits, risks, and opportunities for research and practice. *Journal of Technology in Behavioral Science, 5*(3), 245–257.

Patel, V., Pereira, J., Countinho, L., Fernandes, R., Fernandes, J., & Mann, A. (1998). Poverty, psychological disorder and disability in primary care attenders in Goa. *India the British Journal of Psychiatry, 172*(6), 533–536.

Patel, S. R., & Bakken, S. (2010). Preferences for participation in decision making among ethnically diverse patients with anxiety and depression. *Community Mental Health Journal, 46*(5), 466–473.

Paul, S., Jandhyala, S. K., & Basu, T. (2018, August). Early Detection of Signs of Anorexia and Depression Over Social Media using Effective Machine Learning Frameworks. In *CLEF (Working notes)*.

Pedersen, W. (2008). Does cannabis use lead to depression and suicidal behaviours? A population-based longitudinal study. *Acta Psychiatrica Scandinavica, 118*(5), 395–403.

Pfefferbaum, B., & North, C. S. (2020). Mental health and the Covid-19 pandemic. *New England Journal of Medicine, 383*(6), 510–512.

Radford, A., Wu, J., Child, R., Luan, D., Amodei, D., & Sutskever, I. (2019). Language models are unsupervised multitask learners. *OpenAI Blog, 1*(8), 9.

Rao, S. M., Leo, G. J., Ellington, L., Nauertz, T., Bernardin, L., & Unverzagt, F. (1991). Cognitive dysfunction in multiple sclerosis.: II. Impact on employment and social functioning. *Neurology, 41*(5), 692–696.

Rehman, U., Shahnawaz, M. G., Khan, N. H., Kharshiing, K. D., Khursheed, M., Gupta, K., & Uniyal, R. (2021). Depression, anxiety and stress among Indians in times of Covid-19 lockdown. *Community Mental Health Journal, 57*(1), 42–48.

Resnik, P., Armstrong, W., Claudino, L., Nguyen, T., Nguyen, V. A., & Boyd-Graber, J. (2015). Beyond LDA: exploring supervised topic modeling for depression-related language in Twitter. In *Proceedings of the 2nd Workshop on Computational Linguistics and Clinical Psychology: From Linguistic Signal to Clinical Reality* (pp. 99–107).

Ríssola, E. A., Bahrainian, S. A., & Crestani, F. (2020, July). A Dataset for Research on Depression in Social Media. In *Proceedings of the 28th ACM Conference on User Modeling, Adaptation and Personalization* (pp. 338–342).

Rodrigues Makiuchi, M., Warnita, T., Uto, K., & Shinoda, K. (2019, October). Multimodal fusion of bert-cnn and gated cnn representations for depression detection. In *Proceedings of the 9th International on Audio/Visual Emotion Challenge and Workshop* (pp. 55–63).

Rude, S., Gortner, E. M., & Pennebaker, J. (2004). Language use of depressed and depression-vulnerable college students. *Cognition & Emotion, 18*(8), 1121–1133.

Sanh, V., Debut, L., Chaumond, J., & Wolf, T. (2019). DistilBERT, a distilled version of BERT: smaller, faster, cheaper and lighter. *arXiv preprint* arXiv: 1910.01108.

Shen, G., Jia, J., Nie, L., Feng, F., Zhang, C., Hu, T., ... & Zhu, W. (2017, August). Depression Detection via Harvesting Social Media: A Multimodal Dictionary Learning Solution. In *IJCAI* (pp. 3838–3844).

Siriwardhana, S., Reis, A., Weerasekera, R., & Nanayakkara, S. (2020). Jointly Fine-Tuning "BERT-like" Self Supervised Models to Improve Multimodal Speech Emotion Recognition. *arXiv preprint* arXiv:2008.06682.

Tausczik, Y. R., & Pennebaker, J. W. (2010). The psychological meaning of words: LIWC and computerized text analysis methods. *Journal of Language and Social Psychology, 29*(1), 24–54.

The Conversation. (2020). *Here's how the coronavirus is affecting Canada's labour market.*

Toolan, J. M. (1962). Suicide and suicidal attempts in children and adolescents. *American Journal of Psychiatry, 118*(8), 719–724.

Vaswani, A., Shazeer, N., Parmar, N., Uszkoreit, J., Jones, L., Gomez, A. N., & Polosukhin, I. (2017). Attention is all you need. In *Advances in Neural Information Processing Systems* (pp. 5998–6008).

World Bank. (2020). *The impact of Covid-19 on labor market outcomes: Lessons from past economic crises.*

Wu, X., Lv, S., Zang, L., Han, J., & Hu, S. (2019, June). Conditional Bert contextual augmentation. In *International Conference on Computational Science* (pp. 84–95). Springer, Cham.

Wu, Y., Schuster, M., Chen, Z., Le, Q. V., Norouzi, M., Macherey, W., & Dean, J. (2016). Google's neural machine translation system: Bridging the gap between human and machine translation. *arXiv preprint* arXiv:1609.08144.

Wu, Z., & Ong, D. C. (2020). Context-guided Bert for targeted aspect-based sentiment analysis. *Association for the Advancement of Artificial Intelligence,* 1–9.

Yang, B., Li, J., Wong, D. F., Chao, L. S., Wang, X., & Tu, Z. (2019, July). Context-aware self-attention networks. In *Proceedings of the AAAI Conference on Artificial Intelligence* (Vol. 33, No. 01, pp. 387–394).

Yates, A., Cohan, A., & Goharian, N. (2017). Depression and self-harm risk assessment in online forums. *arXiv preprint* arXiv:1709.01848.

You, Y., Li, J., Hseu, J., Song, X., Demmel, J., & Hsieh, C. J. (2019). Reducing BERT pre-training time from 3 days to 76 minutes. *arXiv preprint* arXiv: 1904.00962.

Zhu, X., & Wu, X. (2004). Class noise vs. attribute noise: A quantitative study. *Artificial intelligence review, 22*(3), 177–210.

Zulfiker, M. S., Kabir, N., Biswas, A. A., Nazneen, T., & Uddin, M. S. (2021). An in-depth analysis of machine learning approaches to predict depression. *Current Research in Behavioral Sciences, 2,* 100044.

Artificial Intelligence in the Energy Transition for Solar Photovoltaic Small and Medium-Sized Enterprises

Malte Schmidt, Stefano Marrone, Dimitris Paraschakis, and Tarry Singh

Abstract The dynamic and rapidly developing European landscape of solar photovoltaic (PV) small and medium-sized enterprises (SMEs) calls for the adoption of artificial intelligence (AI)-based solutions harnessing

M. Schmidt (✉) · S. Marrone · D. Paraschakis · T. Singh
deepkapha AI Research, Assen, The Netherlands
e-mail: malte.schmidt@deepkapha.com

S. Marrone
e-mail: stefano.marrone@deepkapha.com

D. Paraschakis
e-mail: dimitris.paraschakis@deepkapha.com

T. Singh
e-mail: tarry.singh@deepkapha.com

S. Marrone
University of Naples Federico II, Naples, Italy

M. Bertolaso et al. (eds.), *Digital Humanism*,
https://doi.org/10.1007/978-3-030-97054-3_7

105

the power of data. Currently, many SMEs face challenges in putting this approach into practice due to the lack of resources (financial, human, strategic). To aid SMEs in this endeavour, AI maturity assessments have been developed to evaluate the current state of an SME's AI transformation on multiple dimensions along the established maturity stages. However, recommendations on how to advance between the successive maturity stages are fragmented in contemporary literature. In this exploratory study, we conduct thirteen semi-structured interviews and three AI maturity assessments with solar PV plant-operating SMEs in the Netherlands, concluding that the Dutch solar PV industry can be classified as being in the first maturity stage. To transition towards the second stage, our framework emphasizes the need to develop realistic AI strategies, data requirement specifications and human–AI symbiosis. The outcomes of our study may be used by PV energy SMEs as a guide to understand and implement the complex AI maturity stage transitions, as well as by future researchers to build on the proposed framework.

Keywords Artificial intelligence · Solar photovoltaic · Small and medium-sized enterprises · Maturity models

1   Introduction

The global transition to renewable energy sources has become of vital importance in the twenty-first century due to the profound environmental harm caused by exploiting traditional fossil fuels. A popular choice of renewable sources is solar energy, which exists in abundance and harvesting of which has become increasingly accessible through photovoltaics (PV) that generate electricity from sunlight via solar cells (Lupangu & Bansal, 2018). Being the key technology option for a sustainable, decarbonized energy sector, the number of PV installations is expected to increase drastically in the coming years (Jäger-Waldau, 2019).

Unlike conventional, centralized electricity production based on fossil fuels, solar energy capture requires a decentralized approach. This is where small and medium-size enterprises (SMEs[1]) come into play by driving and

---

[1] SMEs are businesses with less than 250 employees and an annual turnover of no more than €50 million.

accelerating PV adoption thanks to their regional knowledge and flexibility (Bacchetti, 2017; Meijer et al., 2019). However, environmental factors such as seasonal and weather-dependent variations add significant complexity to solar energy management. These complexities can be effectively addressed with artificial intelligence (AI) systems capable of generating reliable supply and demand forecasts, as well as a host of other useful tasks based on the available historical and real-time data.

Nevertheless, this adoption can be challenging for many SMEs due to their lack of financial and strategic resources (Matt et al., 2020). In response to the above challenges, AI maturity models have been developed to assess and classify the overall capabilities of an organization against AI maturity stages, which is a crucial first step towards AI adoption (Saari et al., 2019). To date, research studies focusing on AI in the solar energy sector have mainly prioritized technical challenges, leaving to one side other vital aspects including ethical implications, human factors, and organizational adoption by SMEs (Miller, 2014; Verzijlbergh et al., 2016). Veile et al. (2019) conclude that the interplay between technological, organizational and human dimensions represents a promising rationale for incorporating disruptive innovations, such as AI, into business operations.

Currently, there is a lack of holistic guidance for SMEs with respect to transitioning between AI maturity stages, which poses a critical scientific and practical problem statement. We fill this gap through a series of interviews with leading PV plant operators, with the aim of deriving an AI maturity transition framework as an industry-relevant roadmap for embracing a data-driven approach to running a solar PV plant-operating SME. Our paper follows an interdisciplinary approach and considers multiple perspectives to address AI adoption in SMEs. Notwithstanding the global scope of AI-assisted renewable electricity transition in SMEs, our case study focuses on the Netherlands as one of the representative European electricity producers primarily relying on conventional fossil fuels to power a highly dense country and its neighbouring states (Meijer et al., 2019), while displaying a promising and rapidly developing solar industry (Baarsma, 2019).

The rest of the paper is organized as follows. Section 2 provides the necessary background for our study through a literature survey. In Sect. 3, we motivate and detail the methodology followed in our research. In Sect. 4, we present the findings derived from the primary research method and formulate our AI maturity transition framework. In Sect. 5, we discuss managerial implications, study limitations and directions for future research.

2   BACKGROUND

The Dutch Solar Photovoltaic Landscape

The growing interest in solar as a key source for renewable energy in the Netherlands can be traced back to the seminal 1987 report by the World Commission on Environment and Development (Brundtland, 1987; Holden et al., 2014), whose vision for the twenty-first century was to build a global energy ecosystem based on renewable sources in order to alleviate the threats posed by global environmental deterioration. Fast forward to the present day and the annual installed solar PV capacity in the Netherlands has seen a nearly 14,000% increase over the past decade, placing this country among the five leading solar industries in Europe. The potential annual yield of solar electricity in the Netherlands is currently 73% greater than its total electricity consumption (Werther et al., 2019). The fundamental reason behind this growth is the Dutch sustainable energy transition scheme (SDE) which aims to accelerate renewable energy production through subsidies. Much of the emphasis of the SDE has been on solar energy, providing considerable benefits for commercial and utility-scale solar PV companies (Baarsma, 2019), as well as community-owned solar farms.

Despite the remarkable development rate of the Dutch solar PV sector, government targets to address climate change require a much higher share of solar energy than is currently available (McCrone et al., 2020). Another challenge for the Dutch solar energy market is the lack of grid and land capacity for plant operators (Baarsma, 2019). This is exacerbated by significant social non-acceptance of solar plants occupying significant space in often densely populated neighbourhoods. As a remedy for the shortage of land, floating solar projects as well as community-owned solar farms that enhance local ownership have recently become popular (Molengraaf, 2020; Roddis et al., 2020).

Artificial Intelligence for Solar Photovoltaics

The use of AI in solar PV systems entails opportunities for short- and long-term efficiency safety, and reliability of a plant (Jimenez et al., 2017;

Mellit & Kalogirou, 2021). Two specific use cases that stand out are fault detection and solar radiance forecasting.

Fault Detection and Diagnosis
As solar PV systems generally consist of batteries, inverters, converters and PV arrays, various equipment failures may significantly lower the performance and security level of solar PV plants. To this end, AI-driven fault detection and diagnosis (FDD) offers maintenance and operational cost-reduction opportunities to secure consistent overall production (Mellit & Kalogirou, 2021). In general, no less than one-third of the overall production effort is known to be spent on incorrect or unnecessary maintenance (Ahmad et al., 2021), which leaves much room for improvement via FDD.

Commonly, FDD methods consist of three main phases, requiring careful alignment (Fig. 1).

The first phase comprises fault detection in equipment to explore potential anomalies based on the accumulated data. In the second phase, fault classification is performed to identify the cause and nature of the problem. In the third phase, the detected fault is localized and isolated to secure the remainder of the plant (Mellit & Kalogirou, 2021). The first two phases are particularly suitable for machine learning (ML), where

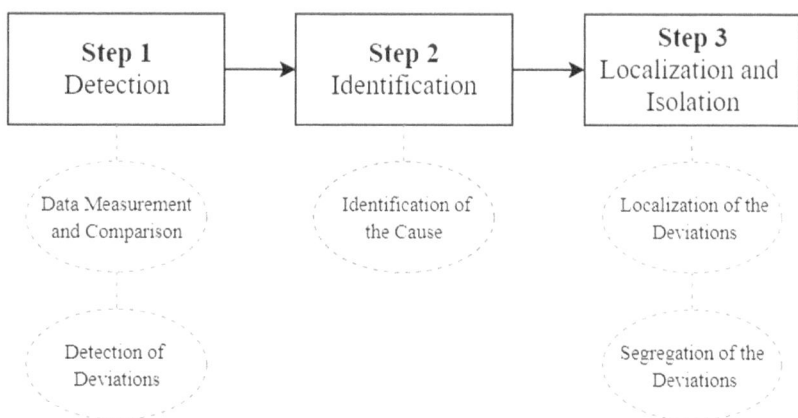

Fig. 1 Three steps of fault detection and diagnosis (Adapted from Mellit and Kalogirou [2021])

classification and *anomaly detection* are among the most common and well-studied use cases. The use of AI thus can help plan and execute maintenance tasks effectively and proactively in a fully automated manner. The high accuracy of models tested for FDD has been shown in recent ML applications achieving a 98.7% classification and identification rate of equipment failure (Madeti & Singh, 2018). It is therefore envisioned that AI-driven smart fault detection systems will become more and more indispensable (Mellit & Kalogirou, 2021), indicating the urgency for SMEs to mature in their AI adoption.

Solar Radiance Forecasting
Predicting radiant energy accurately provides fundamental cost savings and enhanced responsiveness to weather-dependent disruptions that often cause an imbalance in power plants (Kaur et al., 2016). In their work investigating ML-based methods to predict solar radiance, Alzahrani et al. (2017) identified multiple ways to accurately forecast radiance based on data from a solar plant, and concluded immense potential for accurate forecasting. Similar results were reported by Neelamegam and Amirtham (2016), who achieved accuracy rates of 99% in estimating solar energy supply. Such AI applications have great potential for any PV system, making it easier to effectively plan its operation, and evaluate its performance and financial viability (Lorenz et al., 2017).

The highlighted use cases of FDD and solar energy forecasting are important drivers for AI maturity transitions that yield significant benefits for SMEs and will become necessary for competitive advantage (Nishant et al., 2020).

3   CHALLENGES FOR SMALL
AND MEDIUM-SIZED ENTERPRISES

Interdisciplinary adjustments to AI technologies create a "new business logic" (Verhoef et al., 2019), which, especially for SMEs, poses a challenge due to often deficient resources and capabilities. This is especially true of human and financial resources that constrain the complex transition to AI (Matt et al., 2020). Taking the Netherlands as a geographical focus of this study, van Lieshout et al. (Interreg, 2018) conclude that human resource shortages in IT skills and competencies exist primarily in the north of the country. In response to this, the European Commission

(European Commission, 2020) intends to accelerate digital programmes to strengthen collaboration and knowledge sharing between SMEs. Furthermore, increasing financial support is needed as many SMEs are limited in their ability to invest in AI technologies. To counterbalance the lack of finance, the European Commission initiated a pilot investment fund for AI and especially SMEs, the InvestEU programme (European Commission, 2020). However, this is still in its infancy and continues to constrain SMEs compared to large enterprises (Culot et al., 2020).

Artificial Intelligence Maturity Models

With the rise of AI, *maturity models* have been developed to support organizations in their transformation towards disruptive technologies (Alsheibani et al., 2019). More specifically, maturity models serve to (i) understand the current maturity stage in a particular discipline, (ii) define a desired future maturity stage, and (iii) allow for organizational maturity comparison in their respective industry (Saari et al., 2019). The importance of maturity models also refers to solving the ambiguity posed by AI, organizational unawareness regarding AI and the company capabilities required for AI transformations (Alsheibani et al., 2019). Increasingly, scholars and leading technology firms, such as Microsoft, are developing AI maturity models to help conceptualize firms' capabilities along multiple dimensions (Saari et al., 2019). A general theme in relation to AI suggests that "organizations must have established maturity – which encompasses strategy, culture, organizational structures, and core capabilities – to responsibly own an AI-based system" (Charran & Sweetman, 2020).

Maturity models commonly consist of five stages, ranked from the lowest to the highest degree of AI maturity. While the dimensions assessed vary, AI maturity models typically include a data management dimension, a technological infrastructure dimension, a people dimension and a strategic dimension (e.g., Alsheibani et al., 2019; Charran & Sweetman, 2020; Lichtenthaler, 2020). Additionally, maturity models increasingly incorporate ethical, responsible and explainable AI dimensions (Vakkuri et al., 2021).

Recent AI Maturity Assessments

Multiple AI maturity assessments carried out in the recent past indicate that in 2020, AI had not yet made a tangible impact on most businesses (Ramakrishnan, 2020). For instance, Lichtenthaler (2020) claims that most companies still have incomplete knowledge regarding AI while ignoring its opportunities. Similarly, Saari et al. (2019) conclude that "many organizations have only started on the path of exploiting AI in their operations", often facing difficulties in realizing the initial transition steps and typically positioned in the first or second maturity stage (Ramakrishnan, 2020).

AI maturity assessments in the Netherlands echo similar results. For example, recent assessments by the Data Driven Marketing Association in the Netherlands reveal that, on average, companies are in the first stage of AI maturity (DDMA, 2021). To date, detailed AI maturity assessments in the Dutch solar industry are absent. Nevertheless, the empirical study conducted by Meijer et al. (2019), which includes 20 SMEs in the Dutch solar PV industry, reveals that the technological complexity of disruptive technologies, such as AI, is among the most concerning dimensions across the participating firms, which suggests a low AI maturity stage.

Applied Model

The AI maturity model used to assess SMEs in our research is given in Table 1. This model adheres to the required simplicity for SMEs (Schumacher et al., 2016), while providing all core dimensions for an AI maturity assessment. When the chosen model is compared with other scientifically established ones (e.g., Lichtenthaler, 2020; Alsheibani 2019), strong overlaps regarding the dimensions and maturity stages can be found. Furthermore, this model has been applied in multiple industries, adding up to more than 200 successful assessments.

4   Methodology

Motivation and Goal

Two important themes emerge from the previous section. On the one side, a steadily growing solar PV industry in the Netherlands provides a promising outlook for SMEs to drive the sustainable, decentralized electricity transition. In this context, AI seems to be an inevitable tool

Table 1 AI maturity model (Adapted from Ramakrishnan et al. [2020])

Maturity stage/dimension	Stage 1 exploring	Stage 2 experimenting	Stage 3 formalizing	Stage 4 optimizing	Stage 5 transforming
Strategy	Need for AI identified	AI proofs of concept established	Budget for AI increased	AI drives the business	Innovation and transformation
Data	Data requirements are assessed	Data silos are broken and consolidated	Data streams are standardized	Data quality and efficiency properly maintained	Advanced AI techniques targeting all data streams
Technology	Technological infrastructure needed is known	Architecture allows for initial automation	Computing and development tools are streamlined	AI model deployment entirely centralized	Boundaries of technological infrastructure are pushed by innovation
People	First AI interest established by the staff	Interdisciplinary teams and data literacy established	AI community established	Deep AI understanding across the organization	AI solutions complement human resource operations
Governance	Aware of new risks	New risk assessment metrics established	Governance mechanisms are standardized to support technology	AI ethics is embedded in corporate governance	Industry-changing governance mechanisms as best practice example

to stabilize the imbalance and uncertainty in solar PV systems. On the other side, SMEs often lack resources to cope with the complex development of AI and its interdisciplinary impact. Promisingly, AI maturity models provide an initial step for companies to better grasp AI, including the required capabilities and dimensions to work on while identifying a desired future stage. While maturity models serve as an adequate reference point for companies to identify their current positioning regarding AI, a closer look at the literature reveals a fundamental shortcoming (Vakkuri et al., 2021). The core issue of such models relates to the lack of specific action steps to be taken to reach the subsequent stage following a maturity assessment (Santos et al., 2017; Schumacher et al., 2016). To fill this gap, the present study aims to: (i) estimate the current AI maturity stage of Dutch PV operating SMEs, and (ii) identify the key steps for the next-stage AI maturity transition.

Methodological Approach

We conducted thirteen semi-structured interviews using a purposive, non-probability sampling technique (Saunders et al., 2016). The first six in-depth interviews with solar energy consultants and managers of the respective SMEs were used to complement an industry-validated AI maturity assessment of three PV plant-operating SMEs in the Netherlands. This assessment comprises specific questions to identify the maturity stage (Ramakrishnan, 2020). Further, five expert interviews were conducted to obtain the steps required for a maturity transition. The remaining two interviews were used to validate the proposed framework in terms of its feasibility, viability and utility.

5   RESULTS

Maturity Assessment

The results of three maturity assessments (summarized in Fig. 2) place all three companies on maturity stage 1. A common root cause for this AI immaturity, according to many interviewees, is the oligopolist market structure in the electricity industry, resulting in strong interdependence between firms and a high barrier for market entrance, constraining innovation and competitiveness for smaller agents. Besides the market structure, limited AI adoption in SMEs and corresponding low maturity

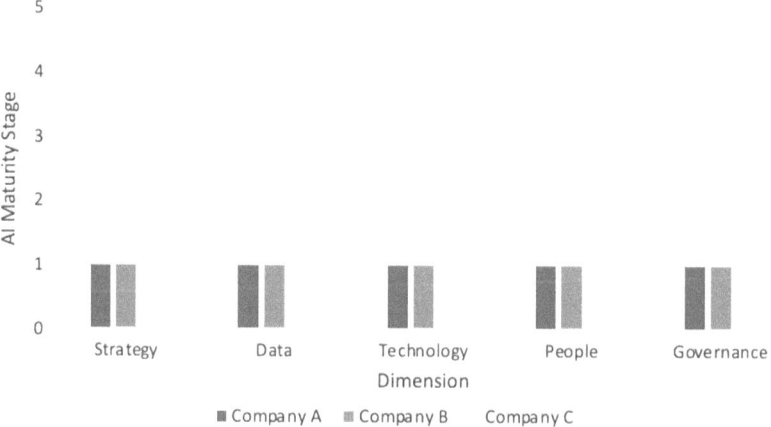

Fig. 2 AI maturity assessment outcomes

levels can be traced to a general lack of awareness and clarity regarding the definition, positioning and usability of AI. On a company level, the reasons for general AI immaturity in SMEs were found to include lack of the right skills and competencies, organizational inertia, lack of an AI strategy, and lack of data management skills. The general consensus from the respondents can be summarized as "in the Netherlands, there is still not a very good framework around sharing data or getting access to that detailed data", as was noted by one of the experts. Nevertheless, a common view among the interviewees was that AI opportunities for solar PV plant operators are multifaceted and inevitable in the long term. Among the opportunities, significant cost reductions for competitive advantage, stabilized grid operations, solar forecasting, enhanced security through fire prevention and predictive maintenance, as well as energy trading were highlighted.

Solution Specification

The outcomes of five expert interviews used to establish the AI maturity transition steps from stage 1 to stage 2 suggest a multidisciplinary alignment between the three crucial dimensions of *AI: strategy, people,* and *data management.*

AI Strategy

The first key dimension towards the second stage of AI maturity is the *strategy*, as confirmed by all interviewees. The overall consensus is that AI has experienced a tremendous increase in attention, often misleading the concrete definition and usability for SMEs. There needs to be an evaluation regarding return on investment when considering AI maturity transitions and the corresponding use cases solved with intelligent solutions. Consequently, the interview respondents insisted that, especially for SMEs, AI projects need to be carefully aligned with the existing capabilities and resources, which often proves a weak point. Taken together, the AI strategy is a fundamental first step to realize an incremental AI maturity transition in the early stages. As AI should solve existing needs or bring proven opportunities, its usefulness must be carefully aligned with the overall vision of SMEs.

Data

Data represents an inevitable pillar for AI solutions and creates a challenge for most SMEs, as a lack of data access and collection reportedly constrains the development of intelligent systems. The respondents commented that one of the key points is to have a clear idea of what kind of data can be obtained. As not all data are accessible in house, external data from third parties is often required. In such cases, data storage and integration of different data streams play a crucial role. On top of that, it was highlighted that integrating data into SME systems is increasingly facilitated through commercial tools, an advantage in often complex data-driven environments. It was also noted that deriving value from data following a thorough data analysis poses a challenge to most SMEs and therefore calls for collaborations with third parties, such as universities or other experts. As a final part of data management, operationalizing data to enable ML operations is a critical element, often challenging for SMEs and requiring in-depth expertise.

People

The pivotal role of *people* in enabling AI solutions was re-emphasized by the interviewees. Among many SMEs, the lack of human resources and technical expertise hampers the realization of AI solutions for businesses. Although workforce–AI alignment displays a crucial intersection, the interviewees agree that external expertise, such as consultants, might be required to initiate a change process in early AI maturity stages.

As reported by the interviewees, collaboration on AI projects between SMEs and universities tends to increase, and provides a reliable source for innovation and "hands-on" expertise. However, some respondents added that training for interdisciplinary projects remains an open issue, which is attributed to the lack of AI-based teaching in universities to facilitate access to talent for employers. However, AI-specific educational programmes on the European level are accelerating, with specific implications for the skills required for the solar sector, which, according to most interviewees, should be a source of talent acquisition for SMEs.

The Proposed Framework

This section presents an AI maturity transition framework for solar PV plant-operating SMEs from stage one to stage two (Fig. 3). The dimensions presented in the previous sections served as a guiding input to construct this interdisciplinary artefact. The framework provides an interrelated and iterative rationale between the suggested steps.

The framework in Fig. 3 was evaluated by two AI maturity stage transition experts, one working for solar PV SMEs in the Netherlands, and one for AI maturity transitions in SMEs globally, including the energy industry. The experts were asked to rate the framework on a scale from 1 (not applicable) to 5 (very applicable) based on three criteria: feasibility, utility and viability (Table 2).

Concerning its feasibility, the experts characterize the framework as a solid, generic tool, covering relevant dimensions and steps to be pursued in early maturity transitions. Regarding the utility of the framework, the experts agree that there is a holistic overview from interdisciplinary dimensions, which poses an indispensable perspective for SMEs to extract the most value from AI adoption. From the viability perspective, the initial version of the framework was criticized for being heavily focused on technical requirements, while missing out other important factors such as user interface, AI–human interaction and data requirements perspectives. This feedback has been reflected in the current framework.

6   CONCLUSIONS

Contrary to previous studies mainly focusing on technical issues when designing AI solutions, the present study pursued an interdisciplinary approach to help SMEs in their AI maturity stage transition. Results of

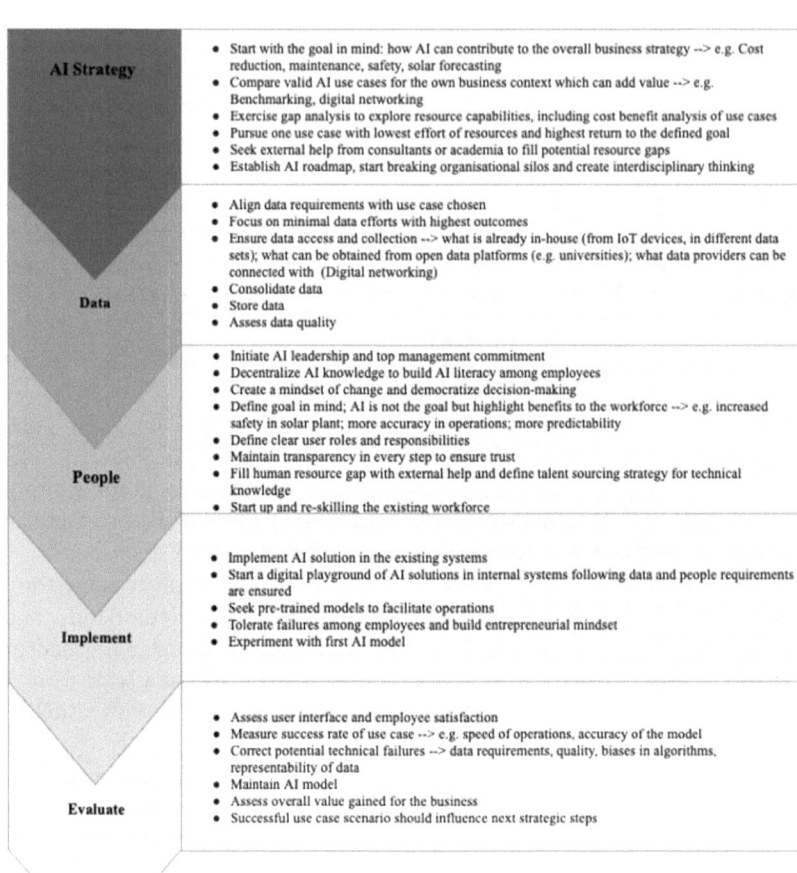

Fig. 3   AI maturity stage transition framework

Table 2   Expert evaluations

Criteria	Expert 1	Expert 2
Feasiblity	4	4
Utility	5	5
Viability	4	3

three maturity assessments and six in-depth interviews assign the current maturity among the studied SMEs to stage one. These findings echo similar results from previous studies in various sectors (e.g., DDMA, 2021; MIT Sloan, 2019). Fundamentally, the lack of resources, such as personnel, to operate AI-based solutions was found to impede maturity transitions at the moment. Additionally, a complex market environment characterized by high levels of regulation and imbalanced competition, considerably hampers AI innovation for SMEs.

Despite the identified immaturity of many SMEs, the present findings confirm that AI will be building the core of the prospective decentralized solar ecosystem (Yang et al., 2020). Digital networking (Verhoef et al., 2019) has been confirmed as a fundamental opportunity to close the often discussed AI resource gap in SMEs. The present findings prove that knowledge sharing between universities and SMEs is accelerating, providing an opportunity for firms with limited knowledge on data management. Another surprising finding of the present study is that building AI leadership and AI literacy in SMEs can positively influence the workforce, breaking organizational silos. These kinds of opportunities for SMEs are not widely addressed in the existing literature.

As the main artefact of this work, a guiding framework for solar PV plant-operating SMEs in the Netherlands was presented and evaluated by two experts, addressing the urgency of closing the gap between scientific advancements in AI, the increasing complexity it reveals and the need for SMEs to adapt to its core logic. Overall, the present study contributes to the emerging topic of AI maturity transitions in solar PV plant-operating SMEs, with both practical and scientific implications.

7   FUTURE RESEARCH

The following suggestions for future research have emerged from the present study. To approach the limited statistical representation of AI maturity assessments in solar PV plant-operating SMEs in the Netherlands, future researchers should pursue such assessments with a larger sample size, following a quantitative research design. As no solar PV-specific AI maturity assessment is currently available, work on such a model is suggested in order to capture the industry needs and capabilities required for SMEs in detail.

In view of the three dimensions identified for AI maturity stage transition—AI strategy, data and people—the remaining maturity dimensions—governance and technology—should also be studied. A similar methodology to that pursued in the present research can be followed. Finally, in view of the lack of guidance for SMEs, the potential of institutionalized support from universities and policymakers to guide SMEs requires further investigation. While the present study proved the existence of such an opportunity, establishing actionable steps needs to come into focus, especially across those dimensions that pose a critical constraint for SMEs—data and people.

REFERENCES

Ahmad, T., Zhang, D., Huang, C., Zha, H., Dai, N., Song, Y., & Chen, H. (2021). Artificial intelligence in sustainable energy industry: Status Quo, challenges and opportunities. *Journal of Cleaner Production, 289.* https://doi.org/10.1016/j.jclepro.2021.125834

Alsheibani, S., Cheung, Y., & Messom, C. (2019). Towards an artificial intelligence maturity model: From science fiction to business facts. *Pacific Asia Conference on Information Systems, 46.*

Alzahrani, A., Shamsi, P., Dagli, C., & Ferdowsi, M. (2017). Solar irradiance forecasting using deep neural networks. *Procedia Computer Science, 114.* https://doi.org/10.1016/j.procs.2017.09.045

Bacchetti, E. (2017). A design approach with methods and tools to support SMEs in designing and implementing Distributed Renewable Energy (DRE) solutions based on Sustainable Product-Service System (S.PSS). *Procedia CIRP, 64,* 229–234. https://doi.org/10.1016/j.procir.2017.03.064

Baarsma, J. (2019). *EU market outlook for solar power 2019–2023.* Retrieved on May 21, 2021, from https://www.solarpowereurope.org/wp-content/upl oads/2019/12/SolarPower-Europe_EU-Market-Outlook-for-Solar-Power-2019-2023_.pdf?cf_id=7181

Brundtland, G. (1987). Report of the World Commission on Environment and Development: Our Common Future. *United Nations General Assembly document A/42/427.*

Charran, E., & Sweetman, S. (2020). *AI maturity and organizations—Understanding AI maturity.* Retrieved on May 4, 2021, from https://query.prod.cms.rt.microsoft.com/cms/api/am/binary/RE4Divg

Culot, G., Nassimbeni, G., Orzes, G., & Sartor, M. (2020). Behind the definition of Industry 4.0: Analysis and open questions. *International Journal of Production Economics, 226.* https://doi.org/10.1016/j.ijpe.2020.107617

DDMA. (2021). *DDMA AI maturity test*. Retrieved on May 5, 2021, from https://aimaturitytest.nl

European Commission. (2020). *On artificial intelligence—A European approach to excellence and trust*. Retrieved on February 3, 2021, from https://ec.europa.eu/info/sites/info/files/commission-white-paper-art ificial-intelligence-feb2020_en.pdf

Holden, E., Linnerud, K., & Banister, D. (2014). Sustainable development: Our *common future* revisited. *Global Environmental Change, 26*, 130–139. https://doi.org/10.1016/j.gloenvcha.2014.04.006

Interreg. (2018). *Industry 4.0—opportunities and challenges for SMEs in the North Sea Region*. Retrieved on February 10, 2021, from https://northsear egion.eu/growin4/desk-study-on-industry-40-in-the-north-sea-region/

Jäger-Waldau, A. (2019). *PV Status Report 2019*. EUR 29938 EN, Publications Office of the European Union. https://doi.org/10.2760/326629

Jimenez, A. A., Munoz, C. Q. G., Marquez, F. P. G., & Zhang, L. (2017). Artificial intelligence for concentrated solar plant maintenance management. *Advances in Intelligent Systems and Computing, 502*, 125–134. https://doi. org/10.1007/978-981-10-1837-4_11

Kaur, A., Nonnenmacher, L., Pedro, H., & Coimbra, C. (2016). Benefits of solar forecasting for energy imbalance markets. *Renewable Energy, 86*, 819–830. https://doi.org/10.1016/j.renene.2015.09.011

Lichtenthaler, U. (2020). Five maturity levels of managing AI: From isolated ignorance to integrated intelligence. *Journal of Innovation Management, 8*. https://doi.org/10.24840/2183

Lorenz, E., Ruiz-Arias, J., & Wilbert, S. (2017). *Forecasting Solar Radiation. NREL Technical Report*. National Renewable Energy Laboratory.

Lupangu, C., & Bansal, R. (2018). Retraction notice to "a review of technical issues on the development of solar photovoltaic systems." *Renewable and Sustainable Energy Reviews, 94*. https://doi.org/10.1016/j.rser.2018.08.016

Madeti, S. R., & Singh, S. N. (2018). Modeling of PV system based on experimental data for fault detection using kNN method. *Solar Energy, 173,*. https://doi.org/10.1016/j.solener.2018.07.038

Matt, D., Modrak, V., & Zsifkovits, H. (2020). *Industry 4.0 for SMEs: Challenges, opportunities and requirements*. Palgrave Macmillan.

McCrone, A., Moslener, U., d'Estais, F., Grüning, C., & Emmerich, M. (2020). *Global trends in renewable energy investment 2020*. Frankfurt School of Finance & Management. Retrieved on May 24, 2021, from https://www. fs-unep-centre.org/wp-content/uploads/2020/06/GTR_2020.pdf

Meijer, L., Huijben, B., Boxstael, A., & Romme, G. (2019). Barriers and drivers for technology commercialization by SMEs in the Dutch sustainable energy sector. *Renewable and Sustainable Energy Reviews, 112*, 114–126. https:// doi.org/10.1016/j.rser.2019.05.050

Mellit, A., & Kalogirou, S. (2021). Artificial intelligence and internet of things to improve efficacy of diagnosis and remote sensing of solar photovoltaic systems: Challenges, recommendations and future directions. *Renewable and Sustainable Energy Reviews, 143.* https://doi.org/10.1016/j.rser.2021.110889

Miller, C. (2014). The ethics of energy transitions. *IEEE International Symposium on Ethics in Science, Technology and Engineering* (pp. 1–5). https://doi.org/10.1109/ETHICS.2014.6893445

MIT Sloan. (2019). *Data, analytics & AI: How trust delivers value.* MIT Sloan Management Review. Retrieved on May 5, 2021, from https://sloanreview.mit.edu/sponsors-content/data-analytics-and-ai-how-trust-delivers-value/

Molengraaf, P. (2020). *EU market outlook for solar power 2020–2024.* Retrieved on May 24, 2021, from https://www.solarpowereurope.org/european-market-outlook-for-solar-power-2020-2024/

Neelamegam, P., & Amirtham, V. A. (2016). Prediction of solar radiation for solar systems by using ANN models with different back propagation algorithms. *Journal of Applied Research and Technology, 14.* https://doi.org/10.1016/j.jart.2016.05.001

Nishant, R., Kennedy, M., & Gorbett, J. (2020). Artificial intelligence for sustainability: Challenges, opportunities and a research agenda. *International Journal of Information Management, 53.* https://doi.org/10.1016/j.ijinfomgt.2020.102104

Ramakrishnan, K. (2020). *The AI maturity framework.* Retrieved on May 27, 2021, from https://www.elementai.com/products/ai-maturity

Roddis, P., Roelich, K., Tran, K., Carver, S., Dallimer, M., & Ziv, G. (2020). What shapes community acceptance of large-scale solar farms? A case study of the UK's first 'nationally significant' solar farm. *Solar Energy, 209,* 235–244. https://doi.org/10.1016/j.solener.2020.08.065

Saari, L., Kuusisto, O., & Pirttikangas, S. (2019). AI maturity web tool helps organisations proceed with AI. *VTT Technical Research Centre of Finland,* 1–7. https://doi.org/10.32040/Whitepaper.2019 AIMaturity.

Santos, C., Mehrsai, A., Barros, A., Araújo, M., & Enrique, A. (2017). Towards Industry 4.0: An overview of European strategic roadmaps. *Procedia Manufacturing, 13,* 972–979. https://doi.org/10.1016/j.promfg.2017.09.093

Saunders, M. N. K., Lewis, P., & Thornhill, A. (2016). *Research methods for business students.* Financial Times/Prentice Hall.

Schumacher, A., Erol, S., & Sihn, W. (2016). A maturity model for assessing Industry 4.0 readiness and maturity of manufacturing enterprises. *Procedia CRIP, 52,* 161–166. https://doi.org/10.1016/j.procir.2016.07.040

Vakkuri, V., Janutnen, M., Halme, E., Kemmel, K. K., Ngyuen-Duc, A., Mikkonen, T., & Abrahamsson, P. (2021). *Time for AI (ethics) maturity model is now.* University of Jyväskylä.

Veile, J., Kiel, D., Müller, J. & Voigt, K.-I. (2019). Lessons learned from Industry 4.0 implementation in the German manufacturing industry. *Journal of Manufacturing Technology Management.* https://doi.org/10.1108/JMTM-08-2018-0270.

Verhoef, P., Broekhuizen, T., Bart, Y., Bhattacharya, A., Dong, J., Fabian, N., & Haenlein, M. (2019). Digital transformation: A multidisciplinary reflection and research agenda. *Journal of Business Research.* https://doi.org/10.1016/j.jbusres.2019.09.022

Verzijlbergh, R. A., Vries, L. J., Dijkema, G. P. J., & Herder, P. (2016). Institutional challenges caused by the integration of renewable energy sources in the European electricity sector. *Renewable and Sustainable Energy Reviews, 75,*. https://doi.org/10.1016/j.rser.2016.11.039

Werther, M., van Hooff, W., Coijn, K. & Blokdijk, R. (2019). *Next-generation solar power: Dutch technology for the solar energy revolution.* Retrieved on May 24, 2021, from https://www.rvo.nl/sites/default/files/2020/08/NL-Solar-Guide-2020.pdf.

Yang, T., Zhao, L., Li, W., & Zomaya, A. (2020). Reinforcement learning in sustainable energy and electric systems: A survey. *Annual Reviews in Control, 49.* https://doi.org/10.1016/j.arcontrol.2020.03.001

PART III

Governance, Regulation and Distribution of Digital Resources

CHAPTER 8

Explorations of the Legal Ethnosphere: Humans, Environment and Robots

Migle Laukyte

Abstract This paper addresses the legal treatment of robots as subjects through the lens of the legal treatment of nature that evolved during the twentieth and twenty-first centuries, when we moved away from an abusive relationship with nature to the recognition that its wellbeing is intrinsically related to our own. In particular, the paper introduces the concept of the legal ethnosphere as a space where legal ideas, symbols, beliefs and insights advance, to elaborate the idea that both nature and robots could evolve from being objects to becoming subjects, thanks to the changes in human understanding of what both—nature and robots— are and what they represent as entities in their own right. The paper focuses on this idea by taking into account the recent legal recognition of rivers by different legal systems in the world and by advancing the idea that similar recognition could be granted to autonomous and intelligent machines in the future. In this context, the paper also advances the idea

M. Laukyte (✉)
Pompeu Fabra University, Barcelona, Spain
e-mail: migle.laukyte@upf.edu

M. Bertolaso et al. (eds.), *Digital Humanism*,
https://doi.org/10.1007/978-3-030-97054-3_8

of autonomy as a boundary object that is being used by different disciplines to shape their objects of study and how, in this particular case of nature and machines, autonomy is becoming a game-changer in our legal consideration of otherness.

Keywords Robot · Human · Nature · Environment · Personhood · Autonomy · Ethnosphere · Law

1   INTRODUCTION

Ethnosphere is an anthropological term used to define the entirety of human ideas, beliefs, convictions, myths, dreams, revelations and insights produced by human imagination since the beginnings of consciousness (Davis, 2007). The term can be easily transferred to the legal domain so as to represent the variety and heterogeneity of legal traditions, ideas and visions that often are transformed into legal regulations and rules, but also frequently remain immaterial proofs that law is more than a sum of norms. This is why this paper refers to the legal ethnosphere as a space where legal theories, ideas, beliefs and images show both our unfolding and our potential for dealing with challenges that we envisage in the future.

It is the future that brings us to the main point of this paper, which is to get to grips with the future of interactions between humans and robots within this web of legal imagination.

In particular, this paper addresses this interaction through the lens of how humans have evolved in treating the environment from the legal point of view. The second part focuses on how the relationship with the environment has changed in modern legal thought in the twentieth and twenty-first centuries, so as to turn, in the third part, to how these changes have impacted on our understanding of nature as a legal person. The goal is to explore how our understanding of personhood in law has changed from being anthropocentric and, thus, centred on humans and human-based entities—such as business companies, political parties, states or nonprofit organizations, which we can always reduce to the people who stand behind them—to become more holistic and include elements of nature, such as rivers. Legal personhood then is no longer related exclusively to humans: there is room to interpret legal personhood as not merely a human condition, but something that could also be attributed to other—non-human—entities.

Finally, in the fourth part, I defend the idea of robots as part of the legal ethnosphere and as a continuation and expansion of a holistic vision of personhood: here, the intelligent, autonomous and interactive robots move away from being objects so as to slowly advance to become in their own right. This conceptual fluidity is a challenging idea, not only in terms of future regulation of machines but also in terms of human ability to learn to co-exist with different forms of life, regardless of whether these forms of life are organic, synthetic or a combination thereof.

Both nature and the robot paradigm are explored in the fifth part through the lens of autonomy as a boundary object that links a variety of different scientific disciplines. The meanings that these disciplines co-produce form the grounds to justify the openness and inclusiveness of autonomy and, consequently, (legal) personhood for non-humans.

2   Law and Nature

It goes without saying that to tell the whole story of the relationship between law and nature is not feasible here. My aim here is more limited: I focus on contemporary ideas about the relationship between law and nature that have become a part of the modern legal ethnosphere.

When I use the term *contemporary*, I refer to the last two centuries—20th and 21st, one finished and the other ongoing—as the timeframe to organize the web of legal imaginary. While the twenty-first century started not that long ago, the 20th was the century of major social, economic, geo-political, legal and cultural changes. We have witnessed what Mumford calls "a radical transformation in the entire human environment, largely as a result of the impact of the mathematical and physical sciences upon technology" (1983, p. 77). How can this transformation be interpreted in terms of the law–nature relationship?

Radical Transformation of the Human Environment: Two Positions

There are two positions by which to interpret this radical transformation. The first is that we have moved away from being inventors of tools to interact with nature and moved towards a "radically different condition, in which [we] will not only have conquered nature but detached [ourselves] completely from organic habitat" (Mumford, 1983, p. 77). This idea is supported by Berry (1999, p. 4), who argues that:

> The deepest cause for devastation is found in a mode of consciousness that has established a radical discontinuity between the human and other modes of being and the bestowal of all rights on the humans. The other-than-human-modes of being are seen as having no rights. They have reality and value only through their use by the human. In this context other-than-human becomes totally vulnerable to exploitation by the human.

A similar perspective is also expressed by Heidegger, who refers to agriculture as a "mechanized food industry", air as a producer of nitrogen, the Earth as a place to yield uranium and even the river Rhine as nothing more than a supplier of hydraulic pressure for the hydroelectric plant (1977, p. 15). This position argues that we have completely subdued nature, that we have found ways to use it for our benefit and nature is dependent on us.

The second position starts from the same premise—that we have moved away from being dependent and subject to the whims of nature—but then reaches a different conclusion: it is not about our detachment from nature but, on the contrary, our re-integration and re-discovery of incorruptible links with nature.

In fact, the twenty-first century seems to suggest the second position has gained ground: the debate about nature and our relationship with it seems to be shifting from the idea of nature as an object of conquest and control to the idea of nature as a living community where humans are but a part of the whole (Berry, 1999). The beginning of the twentieth century left an imprint on the environment because people permitted the "expansion of power, through ruthless human coercion and mechanical organization, [to take] precedence over the nurture and enhancement of life" (Mumford, 1983, p. 77). But we have now realized our errors, those incorrigible effects that our ruthlessness has left, and we do not want to proceed with destroying nature and the environment. We have realized that we are dependent on nature in a variety of ways: our physical and mental wellbeing, our creativity, our food, our spirituality, our leisure, our future, everything depends on nature, on the environment and on the prosperity of the Earth.[1]

We can recognize this shift in our way of thinking by observing how debates, ideas and arguments related to business entities—as the main enemies of the natural environment—have changed: governments, civil society and individual citizens have started to insist on corporate responsibility and accountability for human rights and environmental protection violations that were not heard as often just a few decades ago. In the legal enthosphere, this change in our approach towards corporations and businesses has been reflected by expanding criminal liability to cover not only individuals, but also corporate entities. Such criminal law concepts as guilt, punishment, liability and intention were thus re-interpreted so as to be applicable—even though through legal fictions—to non-human entities, interlacing these concepts with the possibility to "pierce the corporate veil" if need be and to make corporations more socially compliant and aware through the corporate social responsibility framework. It would not be incorrect to make a further argument that it is thanks to our newly established (legal) recognition of nature that we have also deepened our discomfort with the personhood of corporations; indeed, it seems that it is not a new feeling.[2]

Moreover, this shift in our understanding of nature tells us a lot about much bigger shifts and changes that are taking place in our minds, and our cultural and social understanding of where we are and who we want to be: indeed, such broad concepts as autonomy, freedom, control and rights have changed on a more abstract level and the interplay between these abstractions is what makes us comprehend the depth of changes in our understanding of the cosmos.

In the next section, I look at one particular development in the relationship between the law and nature, that is, the emerging recognition of legal personhood for natural phenomena.

---

[1] For instance, Norwegian researchers carried out three studies to assess the effect of plants in the office, in a hospital radiology department and in a junior high school: in all three cases, human health and mental wellbeing were much better with plants nearby (Fjeld 2000).

[2] Hamilton argues that corporate personhood is an illogical yet powerful legal fiction and that "we have never been completely at ease with the corporation as person" (2009, p. 24).

3   LEGAL PERSONHOOD OF NATURE

The previous section briefly sketched the paradigm change in our interaction with nature: Berry (1999, p. 3) defines it as a task "to carry out the transition from a period of human devastation of the Earth to a period when humans would be present in the planet in a mutually beneficial manner".

Indeed, this is one of the biggest lessons that we take away from the 20th-century legal enthnosphere, namely, that nature matters and matters to an extent that it is recognized as a party, a stakeholder, an actor. This means that we no longer perceive nature as an object, but (increasingly so) as a subject. During this period, we have (re)discovered our dependency on the environment and recognized the abuses that we were inflicting (and to a certain extent continue to inflict) on nature, but now there is a new willingness to reduce the negative human impact on nature and stop the destruction before it is too late.

One of the ways to express this willingness is to recognize the legal standing of nature, and in this section, I cite an exemplary case of such recognition: this case involves the Atrato river and the decision of the Constitutional Court of Colombia to recognize it as a subject of rights (Case T-622/16).

What this (and other similar recognitions of legal personhood to other rivers in New Zealand, Canada and other countries[3]) shows is that in the case of nature, legal personhood is also a way to save and protect nature from other humans and corporations.[4] By attributing legal personhood to the Atrato river, the Constitutional Court was elevating the river from being a thing, a resource, an object from which profit is to be made to something closer to humans, that is, to a person. Here, the river is not human in its own right, but something strongly related to humans and

---

[3] In particular, I am referring to the Magpie river in Canada, Wilcabamba river in Ecuador, Whanganui River in New Zealand, Ganges and Yamuna rivers in India, but also to other natural phenomena such as the Te Urewera national park in New Zealand or Lake Erie in Ohio (USA) or general recognition of the environment as it is in the case of the Bolivian law *Ley de Derechos de la Madre Tierra* (Law of the Rights of Mother Earth; *my translation*). Retrieved August 20, 2021, from https://www.bivica.org/file/view/id/2370.

[4] However, the Constitutional Court also wanted to protect humans who lived in the area from mercury and cyanide poisoning, and other toxic materials related to illegal mineral extraction carried out in the area (Sentencia T-622/16).

fundamental to their survival and wellbeing. The health of the Atrato river is the *condicio sine qua non* for the quality of the health of the indigenous people who live by it.[5]

This is how personhood becomes a lifebuoy to make sure that certain natural phenomena will not be destroyed by humans (or human associations in the form of corporations) but, on the contrary, will continue existing because other human (indigenous and Afro-Colombian tribes, citizens) can enjoy and benefit from them. The local communities that live by the Atrato river cannot survive if the river is not "elevated" to become more than just a river: in this sense, we observe how river-object becomes river-subject in law. This is why this process of "personification" became necessary and this is why personhood is a "normative statement about social status" (Hamilton, 2009, p. 18) that the rest of society has to come to terms with.

Undeniably, the legal personhood of rivers brings us back to the idea of legal ethnoscience because of the spiritual and ethnical reasons that have led to granting them legal personhood: in the case of the Whanganui river, it was seen as an "indivisible and living whole and the spiritual ancestor" (Kramm, 2020) of a local Maori tribe and, in the case of the Atrato river, it was seen as a central part of the bio-culture that is generated out of the symbiotic relationship between biodiversity related to the Atrato river and local communities (Case T-622/16).

Hence, the lesson that nature has taught us in this case is re-discovery of categories and re-construction of personhood. In addition, we also learn that nature and its defence call for novel ways to look at the law and its function in this defence. Today, we talk about environmental law, nature protection law, international biodiversity law, biosafety law (Morgera & Razzaque, 2017), wild law (Cullinan 2011) and other branches of law that all recognize the need to establish normative fences that will protect nature from humans and ensure that at least some of the natural phenomena will never again be just objects for humans to exploit.

In the following section, I turn to the personhood of machines as another step towards a new understanding of who or what the person is in law or, at least, what such a person could be in the legal ethnosphere.

---

[5] Compare this vision of a river with the vision of Heidegger when he described the Rhine: it "appears as something at our command" and "... river is dammed up into the power plant; [it]is now ... a water power supplier" (1977, p. 16). I would like to thank Luca Capone for bringing my attention to this example.

4   LEGAL PERSONHOOD AND ROBOTS

In this section, I defend the idea of robots as a part of the legal ethno-sphere and as future examples of the continuation and expansion of a holistic vision of personhood. Indeed, legal debates on personhood are "windows into social values and anxieties at work in law and elsewhere" (Hamilton, 2009, p. 18). In this context, one of these anxieties is repre-sented by artificial intelligence (AI) and robotics and, in particular, by those intelligent, autonomous and interactive robots that move away from being objects so as to slowly advance to become entities in their own right.[6] In Wittgenstein's words, we could talk here about the principle of personification or "the idea that one could beckon a lifeless object to come, just as one would beckon a person" (Wittgenstein, 2018, p. 36).

This conceptual fluidity from robot-object to robot-subject represents a challenging idea not only in terms of future regulation of machines, but also in terms of human ability to learn to co-exist with different forms of life, regardless of whether these forms of life are organic, synthetic or a combination thereof. Indeed, the robot-object to robot-subject flow repeats the flow that nature has already started.

In fact, we have seen the accusation, expressed in Berry's words, that the devastation of nature relies on "radical discontinuity between the human and other modes of being and the bestowal of all rights to humans", whereby these other modes of being have value inasmuch as they are useful to humans (Berry, 1999, p. 4). Are we risking the same accusation in the case of robots?

Obviously, there are many differences between nature and robots in terms of their personhood: one feature in particular stands out: autonomy and how we treat this concept in reference to machines. Technologies of personification are various, and language is the most significant one (Hamilton, 2009). The language—this deposit of mythology (Wittgen-stein, 2018, p. 48)—matters when we have to decide what we are talking about: a thing, a matter, an object, a being, an entity, an individual, a phenomenon, a higher life form, a person, a deity, a living being, an

---

[6] These three features—intelligence, autonomy and interactivity—are features that define the idea of the robot as it is understood in this paper: indeed, there are books on each of these terms, but for the purposes of this work, the point is that these features in machines are understood as features that are similar to how we understand these features when we think about human beings possessing them. I leave the question of consciousness of machines outside the debate. For more on this debate, see, for example, Bartra (2019).

animal, and so on. Hamilton (2009) explains the importance of language in those situations when we turn objects into subjects and vice versa. In the case of robots, we continuously anthropomorphize them, and at the same time, when we do so, and attribute to robots such human characteristics as autonomy, we do not feel comfortable.

In what follows, I will address autonomy in greater detail by looking at it as a boundary object that gains prominence in the legal ethnosphere and connects it to other domains of knowledge.

5   THE QUESTION OF AUTONOMY

Autonomy is a boundary object in the legal ethnosphere that not only connects law to other social sciences, but also opens the door to those scientific domains that are distant from the law, such as computer science, software engineering and robotics.

A boundary object, a concept introduced by Star and Griesemer (1989), is the object that different communities of knowledge use and recognize. It is "at once, stable and plastic or malleable: it is solid enough at its core that the different communities using it will still know they are essentially dealing with the same object, but it is 'soft' around the edges, and in this we can fashion it into different 'shapes' depending on what we are trying to do with it and how" (Laukyte, 2013, p. 227).

Within the realm of the legal ethnosphere, then, the concept of autonomy is such a boundary object that opens the door for new ideas, visions and images coming from other disciplines, such as computer sciences broadly construed. In these sciences, autonomy is a key characteristic that distinguishes a system from an agent: the system as such is not autonomous, whereas the agent is a system with autonomy (Williams, 2010). Furthermore, computer science and software engineering work on more autonomous tools and programs, but at the same time recognize that there is neither framework nor measures that would apply to calculate an object's autonomy with respect to other entities (Williams, 2010).

In the legal ethnosphere, autonomy is no less troublesome: on the one hand, it refers to freedom and independence, the ability be responsible for one's actions and claim an equal status with respect to others. But on the other hand, it also signals the absence of (human or other kind of) control and, therefore, generates uncertainty, risks and the possibility of danger. This explains why, when we focus on AI and robotics, the term "autonomy" itself and the adjective "autonomous" are frequently substituted with other terms: for instance, more often than not, when talking about autonomous vehicles, we talk about automated vehicles and stress that there is no autonomy, but only automation involved.

In the case of nature, the idea of autonomy is framed differently: nature itself is made up of autonomous systems, starting with cells and amoeba and finishing with animals. Furthermore, whole natural ecosystems—coral reefs or rainforests—are autonomous in the sense that they are self-sufficient and exist without any need for human intervention: quite to the contrary, it is the human intervention that more often than not interferes with the flourishing of these systems.[7] Therefore, we can draw a parallel between the idea of autonomy as it could apply to robots and the one that applies to nature: in both cases, the absence of human control and intervention means the freedom of the entity or phenomenon to evolve in its own right and by its own means. This also explains that the essence of autonomy as a boundary object is defined by the human role in its exercise and that we will be able to seriously consider the autonomy of robots, if they show signs of being able to continue their existence without direct support from humans.

The autonomy question triggers many more questions about "power, expectation, control and delegation of responsibility, [and] about whether we can expect our technology to fit in with us, rather than the other way around" (Fry, 2018, p. 149). This is the very point of the autonomy of robots: we are not ready to give it away, and the question is whether we will ever be.

In case we are, Nedelsky offers us a possible solution to our unease with the autonomy of robots: in reference to humans, she argues that "what is essential to the development of autonomy is not protection against inclusion, but constructive relationship" (Nedelsky, 1991, p. 168), and we could borrow this statement to work on our understanding and development of the autonomy of machines.

---

[7] In reference to the idea of self-sufficiency as flourishing, we can also take into account Aristotle's words on self-sufficiency as "that which when isolated makes life desirable and lacking in nothing" (Aristotle 2000, Book I, 7).

6 CONCLUSIONS

The changes in our understanding of nature are not only purely legal and normative but have a deeper impact on our understanding of what we as humans are, and where we stand as beings who challenge their own supremacy and survival through intelligent artefacts and the destruction of nature. From this perspective, then, this paper is a contribution to the never-ending search for settlement between the (continuously expanding) technosphere and (seriously threatened) biosphere.

The legal ethnosphere, as a space of—and for—legal theories, ideas, beliefs and images, represents a space to cultivate and explore the intersections between law and culture. Both are influenced by and produce each other (Hamilton, 2009); both are common grounds and have to respond to the interests of everyone (Gaines, 1991). Furthermore, right now we are "in period of flux where our presuppositions are in doubt, [and] it is therefore possible to exercise some deliberate choice about the frame of reference through which we see the world" (Nedelsky, 1990, p. 184). This period of flux should make us build something new: hopefully, a new kind of relationship with nature, and, hopefully, a new kind of relationship with robots.

In fact, the focus on the personhood of nature and robots is functional to put forward a society we want to live in, and the focus on autonomy also shows the premises that we build the legal personhood on: in both personhoods—that of nature and that of robots— the freedom from humans plays a main role in deciding whether or not a certain phenomenon should be considered autonomous and therefore the subject of rights.

Autonomy remains, within the realm of the legal enthnosphere, a boundary object that has yet to be fully disclosed and enriched with different perspectives and ideas. In fact, there is a doubt whether the existing legal metaphors—personhood, property, etc.—are the right ones on which to build our relationship with nature and, particularly so, with machines. Nedelsky argued 30 years ago that "we need a language of law whose metaphoric structure highlights rather than hides the patterns of relationship its constructs foster and reflect" (1991, p. 163), and perhaps it is more urgent than ever before to follow her advice.

References

Aristotle. (2000) *Nicomachean Ethics.* Retrieved September 23, 2021, from http://classics.mit.edu/Aristotle/nicomachaen.mb.txt.

Bartra, R. (2019). *Chamanes y Robots.* Anagrama.

Berry, T. (1999). *The Great Work: Our Great Way into the Future.* Bell Tower.

Cullinan, C. (2011). *Wild Law: A Manifesto for Earth Justice* (2nd ed). Chelsea Green Publishing.

Davis, W. (2007). *Light at the Edge of the World: A Journey Through the Realm of Vanishing Cultures.* Douglas & McIntyre.

Gaines, J. M. (1991). *Contested Culture: The Image, the Voice, and the Law.* University of North Carolina Press.

Fjeld, T. (2000). The Effect of Interior Planting on Health and Discomfort among Workers and School Children. *HortTechnology, 10*(1), 46–52.

Fry, H. (2018). *Hello World.* Black Swan.

Hamilton, S. H. (2009). *Impersonations: Troubling the Person in Law and Culture.* Toronto University Press.

Heidegger, M. (1977). *The Question Concerning Technology and Other Essays.* Garland Publishing.

Kramm, M. (2020). When A River Becomes a Person. *Journal of Human Development & Capabilities, 21*(4), 307–319.

Laukyte, M. (2013). An Inderdisciplinary Approach to Multi-Agent Systems: Bridging the Gap Between Law and Computer Science. *Informatica e Diritto, Special Issue Law and Computational Social Science, 1,* 223–241.

Morgera, E., & Razzaque, J. (2017). *Biodiversity and Nature Protection Law.* Edward Elgar.

Mumford, L. (1983). Technics and the Nature of Man. In C. Mitcham & R. Mackey (Eds.), *Philosophy and Technology: Readings in the Philosophical Problems of Technology* (pp. 77–85). The Free Press.

Muzyka, K. (2020). The Basic Rules for Coexistence: The Possible Applicability of Metalaw for Human-AGI Relations. *Padalyn: Journal of Behavioral Robotics, 11*(1), 104–117.

Nedelsky, J. (1990, Spring). Law, Boundaries and the Bounded Self. *Representations, Law and the Order of Culture, 30.* 162–189.

Nedelsky, J. (1991). Law, Boundaries, and the Bounded Self. In R. Post (Ed.), *Law and the Order of Culture* (pp. 162–189). University of California Press.

Star, S. L., & Griesemer, J. (1989). Institutional Ecology, 'Translations' and Boundary Objects: Amateurs and Professionals in Berkeley's Museum of Vertebrate Zoology, 1907–39. *Social Studies of Science, 19*(3), 387–420.

Williams, M.A. (2010). Autonomy: Life and Being. In *Proceeedings of the Interantional Conference on Knowledge Science, Engineering and Management* (pp. 137–147). Springer.

Wittgenstein, L. (2018). *The Mythology of Our Language. Remarks on Frazer's Golden Bough*. Hau Books. (Original work published 1967).

Case law.

Sentencia T-622/16 of Colombian Constitutional Court. Retrieved September 4, 2021, from https://www.minambiente.gov.co/images/Atencion_y_part icpacion_al_ciudadano/sentencia_rio_atrato/Sentencia_T-622-16._Rio_Atr ato.pdf

Human Rights in the Digital Era: Technological Evolution and a Return to Natural Law

Aileen Schultz and Mario Di Giulio

Abstract Millions of people worldwide lack a national identity, thus making universal legal identity one of the core targets of Sustainable Development Goal 16: "Peace, justice and strong institutions". With the advancement of digital technologies, we are approaching the possibility of a framework in which each human being can be identified and recognized as a person regardless of their state of origin, and maintain some level of protection independent of any one given nation state or legal

A. Schultz
Data and Model Ethics and Policy, Thomson Reuters, Toronto, ON, Canada
e-mail: a.schultz@worldlegalsummit.org

Founder/President, World Legal Summit, Toronto, ON, Canada

M. Di Giulio (✉)
Pavia e Ansaldo Law Firm, Milan, Italy
e-mail: mario.digiulio@pavia-ansaldo.it

Co-Founder/Vice-President, Thinking Watermill Society, Rome, Italy

© The Author(s), under exclusive license to Springer Nature Switzerland AG 2022
M. Bertolaso et al. (eds.), *Digital Humanism*,
https://doi.org/10.1007/978-3-030-97054-3_9

system. Technologies like decentralized solutions and new-age data governance solutions are leading to the fulfilment of Article 6 of the UN's
Universal Declaration of Human Rights: "everyone has the right to recognition everywhere as a person before the law". In this paper we explore
the global identity challenge and emerging models for new identity and
governance frameworks.

Keyword Human rights · Digital identity · Decentralized technologies ·
Facebook metaverse · Global digital public · Self-sovereign identity ·
Universal Declaration of Human Rights

1   "Who Am I?"

What is the first thing that comes to your mind as you read these words?
Your name.

Our name is a part of our human identity. Above all, our name is
a part of our human dignity. The book *Les Misérables* by Victor Hugo
illustrates this through the character of Jean Valjean, who was referred to
as 'Prisoner 24601' before escaping and creating a life in search of the
dignity that was taken from him by the state as a nameless prisoner. In
addition, a name is not only a name because it colours the identity of
a person. After all, as Shakespeare has so beautifully illustrated, "What's
in a name? That which we call a rose by any other name would smell as
sweet" (Shakespeare, 1993).

The achievement of a legal identity for all is one of the Sustainable Development Goals (SDGs) agreed to by all UN (United Nations)
member states. Recognition as a legal person is fundamental in order to be
a recipient of public services, and without this legal recognition a person
may be denied access to rights such as health care or the exercise of civil
and political rights. Further, the need for recognition as a legal entity
is not limited to the identification of an individual, but is required also
for the acknowledgement and record of most life events, such as births,
deaths, marriages and divorces (Mrkić, 2019).

The address of human rights issues is foundational to the construction of the SDGs. The United Nations Declaration of Human Rights is
acknowledged by 193 member states, and is supranational in its application in that it is intended to apply to each human being regardless

of origin of state (United Nations, 1948). While the theory of natural law is ripe for debate, the fact that the Declaration stems from the UN's founding charter and has been adopted by member nations through the General Assembly would suggest that we can at some level put into words those rights that we universally accept to be naturally granted to each human being at birth. In this respect the Declaration might be referenced, at least in part, as a statement of natural law, for which we ought to seek positive law enforceability, as has been suggested by Jacques Maritain, as referenced within the UN framework (Maritain, 2018):

> We then understand how an ideal order, with its roots in the nature of man and of human society, can impose moral requirements universally valid in the world of experience, of history and of facts, and can lay down, alike for the conscience and for the written law, the permanent principle and the primal and universal norms of right and duty.

However, Article 6, "everyone has the right to recognition everywhere as a person before the law", is in fact a paradox with little real-world tangibility. Today, each human person is recognized before the law of a given nation, whether by citizenship or form of residency, or is otherwise "stateless" in that they do not have a given nationality. Millions of people worldwide are currently stateless, and as such their personhood and human rights as recognized and protected within an enforceable positive law framework may be compromised. Therein lies the paradox.

With the advancement of digital technologies, we are approaching the possibility of a framework in which each human being can indeed be identified and recognized as a person regardless of their state of origin and maintain some level of protection independent of any one given nation state or legal system. There is a growing tension between, on the one hand, the universality of technological tools and human rights, including the right to a legal personality, and, on the other, the concept of state sovereignty and prerogatives, which risk being increasingly eroded by the development of new digital tools. This tension is ever increasing with the advancements of digital capabilities and the inherent universality of these technological applications. It is this tension that we endeavour to explore.

2   THE GLOBAL IDENTITY CRISIS

According to the UN Refugee Agency, millions of people worldwide lack a national identity, thus making universal legal identity one of the core targets of Sustainable Development Goal 16: "Peace, justice and strong institutions". The road to accomplishing this is not easy and, while technologies may help, the cooperation of national authorities and individual states is crucial for the implementation of systems as well as for the conferment of legal effect.

Traditional analogue identity frameworks are in effect in every jurisdiction today. These are physically based forms of identification issued by government bodies and maintained by government institutions, and are often further legitimized through the issuing of bank accounts, academic credentials or health records. There are several issues with these traditional frameworks, such as their susceptibility to identity fraud, non-transferability amongst states, and costly error-prone infrastructure. These issues have created demand for new identity management mechanisms, such as digital solutions can provide.

On the topic of digital solutions, computer scientist and modern-day philosopher Luciano Floridi explains that we are living in a time where our life switches continuously between being online and in real life (IRL), where we are no longer operating in one or the other but are rather operating in a hybrid function that he calls "onlife" (Floridi, 2017).

In real life we are accustomed to ID documents and paper certificates; however our use of digital instruments is increasing exponentially in a way that would have been very difficult for us to foresee just 50 years ago. For example, we might consider the growing use of crypto-currencies or the creation of virtual worlds where people purchase digital assets that are for use strictly within the virtual environment, such as what we might see in Facebook's *Metaverse*.

This is also the case for digital identities, the use and adoption of which are becoming more mainstream, providing alternative mechanisms for the issuance and management of identity.

3 IDENTITY INFRASTRUCTURE FOR THE DIGITAL AGE

On Wikipedia a digital identity is defined as "information on an entity used by computer systems to represent an external agent".[1]

Technological advancement and increasing demand from citizens for improved government systems have led to the widespread adoption of digital identity systems by governments worldwide. According to a 2016 report by the World Bank (World Bank, 2016), most developing nations had some form of digital identity framework in place. These frameworks range in levels of complexity, starting with very basic digital ID cards that facilitate identification purposes, to the use of digital IDs that are biometrically authenticated and enable secure, private and streamlined access to a wide range of government services.

It is these more advanced capabilities that are enabling the path toward new identity frameworks that enhance the individual's ability to maintain control and management of their identity and related data, while at the same time enabling protected social involvement within a digital environment. We will now explore the Estonian use case.

E-Estonia, Mater Omnium

Although Sierra Leone is reportedly the world's first "smart country"—a nationwide identification programme based in digital credentials and blockchain was used by the electorate in 2017 (McKenzie, 2018)—Estonia has nonetheless been recognized as the poster child of e-government platforms and new-age identity systems since as far back as at least 2016 (Reynolds, 2016).

Estonian e-government services span across "virtually all state-related operations except marriage, divorce, and real estate transactions" (Paadam & Martinson, 2019). Here, however, we will focus on their pioneering identity infrastructure. Indeed, it is worth noting that the e-ID developers behind the Estonian digital identity platform have been recognized by a prominent group of organizations including Google, the United Nations, the European Union, and the US State Department (Paraskevopoulos, 2021).

---

[1] This definition is intentionally taken from Wikipedia to illustrate just how mainstream the concept has become.

Foundational to the Estonian e-government platform is a new-age, decentralized and AI-enhanced identity management infrastructure, along with a web of laws or policies designed to protect citizens while facilitating the government's use of digitally enabled services (Paadam & Martinson, 2019).

Although it has been argued that e-Estonia is not technically blockchain enabled, the Estonian e-governance framework does operate on a platform which enables connectivity and communication between local servers without centralized storage. X-Roads is the platform the Estonian government operates on, and beyond connectivity between disparate systems it is also a tool to write to varying information systems, transfer data between them and enable mass cross-system searches (Cullell, 2019).

Most compelling about this infrastructure is perhaps its adoption by other states. Most notably, in 2019 Finland and Estonia began efforts to connect their business registers. Assuming successful implementation, this would enable an unprecedented digital interoperability between states that would enhance efficiency, data quality and the cost-effectiveness of operations digitally and internationally (Plantera, 2019).

Critical to the cross-functionality of services that X-Roads enables is the Estonian identity scheme, which is designed to ensure secure and reliable identification of natural and legal persons while creating cohesiveness between physical and digital identity management systems. At the core of the Estonian digital ID agenda is ensuring universal utility, where one person should have one source of identification that permits use of government services in an interoperable and convenient way.

Although one of the primary principles driving the Estonian agenda is "to have as little state as possible, but as much as is necessary" according to the government website, there are still centralized frameworks driving much of the data exchange and governance of identity systems. Truly new-age frameworks, on the other hand, are offering solutions for maintaining the benefits of traditional state structures, while removing dependency on these structures for social, economic and political participation.

4   Our Globally Connected Digital Public

The Internet has opened new frontiers; indeed, it has effectively eliminated jurisdiction-bound borders. Our global digital public is becoming more and more integrated through the harmonization and popular preference of platforms, frameworks, protocols and standards. We need only look at any given social media post, produced by any given person on this planet to see just how globally permeable and socially intertwined our lives have become.

For example, consider the global market share of Google, reported to be 86–92% in 2021 (Johnson, 2021a, 2021b). Looked at another way, that is effectively over 4 billion of the 4.66 billion Internet users worldwide using Google as their primary search engine (Johnson, 2021a, 2021b). This means that Google alone is the primary information provider for the entire Internet-using world, making its search algorithms the effective standard by which information is distributed and made accessible online.

Taking another well-known example, let's look at Facebook, with nearly 3 billion active users each month as of mid-2021—that's nearly 75% of the Internet-using population actively engaged on the platform regularly (Statista Research Department, 2021a, 2021b). In addition, WhatsApp (owned by Facebook) is the most popular mobile messaging app, with roughly 2 billion active monthly users (Statista Research Department, 2021a, 2021b).

This increasing homogenization of Internet-based services and their use is certainly cause for great concern, and is a driver behind the ongoing efforts by organizations like the World Wide Web Consortium (W3C), Institute of Electrical and Electronics Engineers (IEEE) and the International Organization for Standardization (ISO) to come up with international solutions for governing some of the underlying systems and technologies making these platforms possible.

However, these trends are also supporting the feasibility of, and opportunity to envision, a global environment in which we can operate as individuals, alongside others from all over the planet, without being bound by locality and centralized institutions like banks or governments for economic, social or even political participation. However, to arrive at true global utility that is sustainable in the long term we are in need of more advanced identity solutions.

5   Technological Evolution and the Road Ahead

We have reviewed the current digital identity landscape, and have distilled what can be considered as our *global digital public*, a digital connectedness amongst nearly 50% of the world's population enabled by growing accessibility to Internet infrastructure, products and services. We will now take a closer look at a couple of core technological advancements that are making possible long-term identity solutions fit for this digital public. In particular, we will look at decentralized and distributed technologies, and data accessibility, management and self-governance models.

Decentralized and Distributed Infrastructure

Decentralized and distributed systems are not actually new. Distributed databases, for example, came into mainstream adoption with the World Wide Web, the well-known distributed information system much of the world uses today (Winstead & Gershon, 2013), and which is enabled by the Internet, a globally distributed system of interconnected computer networks that facilitates communications between devices and other networks. The Internet alone has given way to countless socially impactful use cases for decentralized and distributed systems.

For example, today entire economies of stateless or invisible populations (people without any form of government-issued identification) can transact through the Internet without a requirement for localized identification or credentials. This is made possible through the use of crypto-currencies and applications like M-PESA, a money transfer service that operates entirely through users' mobile devices. M-PESA is considered to be a "branchless banking service" where users can deposit and withdraw money without the need for a bank account (Warwick, 2021).

Furthermore, decentralized computing has been in use since the 1980s, with for example the application of concepts introduced by David Chaum who patented the first payment system to operate through a decentralized computing mechanism (Chaum, 1982). In Chaum's work it is already clear just how integral identity was understood to be in the real-world application of his concepts, with his patent application called "Cryptographic identification, financial transaction, and credential device". The growing popularity today of blockchain-based infrastructure for identity systems, is testament to the legacy of Chaum's work. In the patent filing we find the following excerpt:

The apparatus may be utilized by its owner to identify himself to an external computer system ... The apparatus, in one embodiment, is separable into a cryptographic device, packaged in a tamper resistant housing, and a personal terminal device. The cryptographic device includes ... a memory device for storage of data necessary to allow identification of the owner, and control logic for controlling the exchange of data with the external system to identify the owner...the decryption of this data requires the entry of a secret ID, known to the owner. (Chaum, 1985)

This excerpt is a near perfect explanation of the mechanism by which today's blockchain- based crypto-economies operate from the user's end. Today crypto-currencies are made possible through digital wallets stored on personal devices (phones and computers) that have associated private keys/addresses or "secret IDs" that interface with external systems, for example the e-commerce systems that accept crypto-currency payment (Antonopoulos, 2017).

In an identity management system this secret ID mechanism enables people to initiate their identity or membership in a particular network via a private key that only they can use, with for example, biometric authentication or another private pairing key associated only with their device. Provenance of the person's identity thus lies with themselves as opposed to a third party like a government or other identity-issuing institution. This identity mechanism then enables the person to transact or otherwise participate online privately and anonymously, and with complete control and ownership over their own identity.

Data Accessibility, Management and Self-Governance Models

It is estimated that by 2025 connected populations will produce 463 exabytes of data each day worldwide; to put it into perspective, that's 1,000,000,000,000,000,000,000 bytes of data a day (Desjardins, 2019). It's worth noting that those are pre-COVID-19 pandemic estimates. An infographic produced in 2019 by the Raconteur publishing house (Raconteur, 2019) paints a vivid story of what that digital data footprint might look like on any given day; on average, 500 million Tweets are sent, 300 billion emails are sent, 4 terabytes of data are produced by connected cars, and 30 petabytes of data are produced by wearable devices.

In aggregate, this content and information about us can create access to extremely precise portfolios about who we are, what we like and dislike, what motivates us, what makes us emotionally unstable, and can be used by marketers, institutions and other interested parties in a number of different ways. It can be used for example to manipulate our political leanings as we saw with the 2018 Cambridge Analytica scandal, or to target us online with ads that are tailored to our most pressing needs and desires. This information is also sold in unregulated markets and the dark net and can be used for more nefarious things like identity fraud, impersonation or blackmail (Bartlett, 2015).

Further, there are increasing concerns about government use and protection of citizens' personal data. According to a report entitled the 'AI Global Surveillance (AIGS) Index', at least 42% of the 176 nations surveyed make use of AI surveillance technologies to gain information about their citizens. One of the primary technologies used by these nations is facial recognition tools that make use of image data about people for various purposes, including informing local policies, policing or the development of smart-city infrastructure (Feldstein, 2019).

This is all data about people that often does not belong in any one *jurisdiction* or within any one governance framework or body of regulatory protection. We are thus in dire need of new models for managing and protecting this information. These factors have led to increasing investment in building data governance frameworks and tools that not only support the protection and management of personal information, but that also provide mechanisms for self-sovereignty and control within a global context. Here are a few core examples.

Verifiable Credentials Data Model 1.0
The W3C is working on global standardization models for verifiable credentials, intended to support the verifiability of information pertaining to us on the Web. It is currently very difficult to determine the authenticity of physical credentials such as diplomas and professional certificates, or government-issued IDs like passports and driving licences. Furthermore, these currently centralized forms of personal information are not interoperable and are often issued by third parties, making it complicated and burdensome to use them in a digital or Web-based environment.

One of the primary components of the verifiable credential model is the decentralized identifier (DID). The DID is a portable URL-based identifier that is affiliated with something that has an individual existence, such as an organization or person. The verifiable credential is a string of characters that incorporates the identifier. This allows for the portability and immutability of a person's ID and associated credentials, without reliance on the issuing body for reissuance or verification (Sporny et al., 2019).

MIT's Enigma Project

In *Trusted Data*, the authors express the following core issues with personal data governance models: first, the need for the creation of mechanisms for people to have control of their personal data; and second, the infrastructure to track and manage this information in terms of which systems (or algorithms) are making use of this data (Shrier, 2019). These issues serve as much of the motivation behind the proposal and growing adoption of the concepts disclosed in the paper titled, "Enigma: Decentralized Computation Platform with Guaranteed Privacy" (Zyskind et al., 2021).

Enigma is a peer-to-peer network or platform that allows for multiple parties to collectively make use of data while maintaining complete privacy. The network is controlled by an external blockchain-based infrastructure which enables a mechanism by which to control and manage use of the data in the distributed, decentralized fashion we detail in previous sections (Zyskind et al., 2021). This platform enables the use of personal data by third parties in such a way as to allow people to transfer and utilize personal data safely and in a way that is immutable and secure, thereby ensuring trust in the system (Shrier, 2019).

The Sovrin Network

> Despite a quarter-century of advances in Internet technology, there is still no easy way to prove online that you are not a dog, are over 18, live at a certain address, graduated from a certain school, work at a specific company or own a specific asset.

This is a statement found in a White Paper titled 'Sovrin: A Protocol for Self Sovereign Identity and Decentralized Trust', published by the Sovrin Foundation, a non-profit organization that provides the operation and infrastructure of the Sovrin Network. The Sovrin Network is a blockchain-based digital identity system designed to build a proof of identity fit for the global digital age. The network makes use of open-source infrastructure and standardization, like the decentralized identifier being built by the W3C, to support its institution and operation.

The network is founded on the principle of becoming a global public utility, meaning that the system must operate by the same open code and governance models as today's Internet and Web protocols. By this token the network can be used by anyone and is not owned by any one organization or person, but rather is completely decentralized and operates through a membership-based nodal infrastructure called the Sovrin Trust Framework. It is principally designed to provide a trust-based, globally operable framework for identity management that enables individuals to have control over their own identity and does not rely on any one central institution or organization to provide verification.

6   Next Generation Digital Publics

The combination of technological advancements detailed above, along with the growing global pressure for improved government services, less corruption and security within a digitally global context, has led to the emergence of new-age identity models.

As a leading example, the uptick in crypto-currency use in the last decade is a pretty strong indicator of the traction non-state-based models for Internet- based societies can have. Another great example comes from Catalonia, Spain, where a separatist movement was facilitated by a blockchain-based citizen platform designed explicitly to offer alternative government participation (Lomas, 2019).

Perhaps one of the most extreme examples is Freeland. While much of the mainstream content about Freeland is now (perhaps ironically) inaccessible or corrupted, there are still sufficient remnants available to showcase just how much demand there is for alternative frameworks for our digital public. Freeland is an experiment for a virtually based, digitally enhanced, stateless society that would offer much of the same services one might receive from any given government (Safronov, 2021). In its short life-span it included much of the technical and philosophical components of what a truly stateless society might look like.

This growing demand for alternative, digitally based identity and governance models has led to institutionalized models for achieving some of the same ends. Most prominently, the European Commission has initiated plans for a robust, interoperable, e-government system designed to enhance digital public services for Europeans (European Commission, 2021). Core to this initiative is the design and adoption of a common platform for identity management that is electronically based. One of the primary drivers for this model is the need for cross-border interoperability and the access to public services while abroad.

Another example of cooperation amongst states can be found with the ID4Africa initiative, which is an NGO designed to support nations across the continent with the adoption of digital identity frameworks and cooperation amongst states. ID4Africa is focused on enabling digital economies and interoperability amongst nations, ensuring everyone has a legal identity and access to human rights (Vyjayanti et al., 2018).

7 Challenges and Considerations

We are living through an incredible period of history from the late twentieth to the early twenty-first century, when the progress of digital technologies is allowing humanity to visualize a future of wellness and peace. However, and rather unfortunately, two world wars and endless domestic conflicts have taught us that technologies can simultaneously boost human activities up to the sky but also straight to a proverbial hell.

While there are numerous issues that will need solutions if we are to arrive at a sustainable model for universal identity, we will here focus on the ethical use and governance of personal data.

Protection of Personal Data

One of the greatest challenges with adopting digital identity frameworks is the ethical use and protection of personal data. There are well-founded concerns related to the risk that electronic data associated with a person may be used beyond the intended purposes, where for example information used for setting up an account on an online platform may be illegally acquired by hackers and criminals, or manipulated and used for non-consensual purposes by the platform itself (Di Joe, 2021).

As the mechanisms by which identity is established evolve, with for example the use of biometric data for registration and authentication or decentralized technologies, so do the risks and challenges. There are growing sensitivities in particular around the use (or misuse) of personally identifiable information (PII), of which biometric data is the most valid form. Given the immutability of biometric data, there has been widespread adoption by governments mandating its use in the management of government-issued identity programmes (Accenture, 2017).

It seems that many of the concerns with the use of PII are related to the lack of predictability and the distrust surrounding its use by digital platforms and institutions. For example, Germany's mandatory biometric (fingerprint) registration for government-issued identification began with a voluntary citizen adoption period, during which only 15% of applicants indicated acceptance of the programme. Some of the main issues were said to be scepticism and fear of institutionalized corruption (Werkhäuser & Esther, 2021).

During a conference held in Madrid by the Thinking Watermill Society on November 15, 2019,[2] the audience expressed strong aversion to facial recognition systems, on the grounds that it was unclear how their data would be used and protected.

Institutional trust is integral to the wider adoption of identity systems with universal utility. Since the European Union's General Data Protection Regulation (GDPR) came into effect in 2018, there has been paramount activity internationally with regard to building localized standards and processes for data protection. However, when it comes to the practical implementation and governance of regulatory structures, systems are lagging behind, leaving many organizations and digital platforms to self-regulate. For example, there are no universally accepted standards or regulations for the implementation of AI systems, systems that are developed only with access to large amounts of data. In early 2021 the EU released what is being called the first ever draft regulation for AI, the foundational driver of which is the widespread development and adoption of so-called trustworthy AI (European Commission, COM, 2021).

Instability in Governance and Infrastructure

The Italian saying "the country you go, the custom you find" is very true even when applied to our virtual environments. For example, according to Jeff Horwitz, a reporter for the *Wall Street Journal*, Facebook has a "whitelist" of users that are exempt from the rules of use intended to protect other users and how information is shared on its platform (Horwitz, 2021).

---

[2] Café La Fabric (Madrid), November 15, 2019. Artificial Intelligence: concrete applications in biometric identification. Ethical concerns and legal protection in an EU perspective.

Although many countries have adopted data regulation frameworks and digital identity infrastructure, these structures can be unstable and inconsistent, leading to system failure in the long term and a reinforcement of mistrust. For example, the World Legal Summit, 2019 carried out an analysis of government readiness for decentralized identity systems, indexing variables such as the instalment of digital identity infrastructure paired with regulatory instruments for the protection of persons' data. It was revealed that several nations have active digital identity programmes in place, without the necessary data protections.

Particularly worrying is the case of revolutions, *coups d'état* or government overthrows, where new groups coming to power may exploit existing vulnerabilities. We saw this situation in mid-2021 as the Taliban swept across Afghanistan, regaining power after nearly two decades of US military presence across the country. The US military used biometric information to identify and work with members of the Afghan–US coalition. There is now concern that the Taliban may have access to this data and can target collaborators (Guo & Nooriarchive, 2021).

It is worth noting that traditional analogue systems are susceptible to many of the risks described above. Furthermore, many of the technical advancements detailed in a previous section are enabling digital capabilities designed explicitly to better protect data and prevent abuses by governing bodies. Nevertheless, the idea of a global identity system and utility raises the fear of a universalized Big Brother (Kuiper, 2016), and fundamental protections and controls need to be developed.

8   Conclusion

Further consideration of these novel emerging topics, some aspects of which are not yet fully understood, requires an interdisciplinary approach across computer sciences, legal studies, political science, sociology, and newer disciplines such as future studies, human–computer interaction (HCI) and user-centred design. There are also a number of challenges that cannot be ignored.

We have shown the demand for inclusive identity solutions that recognize and protect the natural rights of all human persons regardless of state of origin, and have illustrated the emerging technical capabilities that are making such a reality possible. However, until there is a willingness by local authorities to implement policies and digital instruments for the achievement and protection of a legal identity for all, the realization of such a system will remain out of reach.

Nonetheless, it is clear that the necessary technological tools and systems are about to become mainstream, and are already being adopted in pockets across the globe, making feasible a framework in which the human person can be recognized and protected solely based upon their natural origin as a human being. In sum, it is quite possible that in our life-time we will bear witness to the universal manifestation of the UN Declaration of Human Rights and Jacques Maritain's "ideal order", where the union of man, state and, in our case, technology, will harmoniously emerge.

References

Accenture, J. V. (2017, August). *Biometrics and advanced analytics revolutionizing how governments address data security and privacy, Accenture report finds.* Retrieved October 4, 2021, from https://newsroom.accenture.com/news/biometrics-and-advanced-analytics-revolutionizing-how-governments-address-data-security-and-privacy-accenture-report-finds.htm

Antonopoulos, A. M. (2017). Wallets. In *Mastering bitcoin: Programming the open blockchain: Unlocking digital cryptocurrencies.* O'Reilly & Associates Inc.

Baker McKenzie. (2018, November). *Blockchain and cryptocurrency in Africa.* Retrieved September 15, 2021, from https://www.bakermckenzie.com/-/media/files/insight/publications/2019/02/report_blockchainandcryptocu rrencyreg_feb2019.pdf

Bartlett, J. (2015). *The Dark Net: Inside the Digital Underworld.* (n.p.).

Chaum, D. (1985, July). *Cryptographic identification, financial transaction, and credential device.* Retrieved September 11, 2021, from https://patents.goo gle.com/patent/US4529870

Chaum, D. L. (1982). *Established, maintained, and trusted by mutually suspicious groups.* Retrieved September 11, 2021, from https://chaum.com/publicati ons/research_chaum_2.pdf

Cullell, L. M. (2019, October). *Is e-Estonia built on blockchain technologies?* [Fact Check] Retrieved September 28, 2021, from https://hackernoon.com/e-est onia-is-not-on-blockchain-22iy2gx6

Desjardins, J. (2019, April). *How much data is generated each day?* Retrieved August 18, 2021, from https://www.weforum.org/agenda/2019/04/how-much-data-is-generated-each-day-cf4bddf29f/

Di Joe, T. (2021, July). *How your personal data is being scraped from social media.* Retrieved September 7, 2021, from https://www.bbc.com/news/bus iness-57841239

European Commission, COM(2021) 206. (2021, April). *Regulation of the European parliament and of the council laying down harmonised rules on artificial intelligence (artificial intelligence act) and amending certain union legislative acts*. Retrieved October 1, 2021, from https://eur-lex.europa.eu/legal-content/EN/TXT/?uri=CELEX%3A52021PC0206

European Commission. (2021, June). *eGovernment and digital public services*. Retrieved October 3, 2021, from https://digital-strategy.ec.europa.eu/en/policies/egovernment

Feldstein, S. (2019, September). *The global expansion of AI surveillance*. Retrieved August 18, 2021, from https://carnegieendowment.org/2019/09/17/global-expansion-of-ai-surveillance-pub-79847

Floridi, L. (2017, January). *We are neither online nor offline, but onlife*. Retrieved September 7, 2021, from https://www.chalmers.se/en/areas-of-advance/ict/news/Pages/Luciano-Floridi.aspx

Guo, E., & Nooriarchive, H. (2021). *This is the real story of the Afghan biometric databases abandoned to the Taliban*. Retrieved October 1, 2021, from https://www.technologyreview.com/2021/08/30/1033941/afghanistan-biometric-databases-us-military-40-data-points/

Horwitz, J. (2021, September). *Facebook says its rules apply to all*. Company documents reveal a secret Elite that's exempt. Retrieved September 7, 2021, from The Facebook Files—WSJ.

Johnson, J. (2021a, July). Global market share of search engines 2010–2021a. In *Statista Worldwide desktop market share of leading search engines from January 2010 to June 2021*. Retrieved July 30, 2021, from https://www.statista.com/statistics/216573/worldwide-market-share-of-search-engines/

Johnson, J. (2021b, September). Worldwide digital population as of January 2021b. In *Statista Global digital population as of January 2021*. Retrieved July 30, 2021, from https://otterbein.libanswers.com/faq/153488

Kuiper, K. (2016). *Big brother fictional character*. Retrieved October 1, 2021, from https://www.britannica.com/topic/Big-Brother-fictional-character

Lomas, N. (2019, October). *Catalan separatists have tooled up with a decentralized app for civil disobedience*. Retrieved October 3, 2021, from https://techcrunch.com/2019/10/17/catalan-separatists-have-tooled-up-with-a-decentralized-app-for-civil-disobedience/

Maritain, J. (2018). *Human rights and natural law*. UNESCO Courier 2018-4. Retrieved October 14, 2021, from https://en.unesco.org/courier/2018-4/human-rights-and-natural-law

Mrkić, S. (2019, June). *United Nations strategy for legal identity for all concept note developed by the United Nations legal identity expert group*. Retrieved September 7, 2021, from https://unstats.un.org/legal-identity-agenda/documents/UN-Strategy-for-LIA.pdf

Paadam, K. J., & Martinson, P. (2019). E-Government & e-justice: Digitizations of registers, IDs and justice procedures. In M. De Stefano, G. Dobrauez, & S. Verlag (Eds.), *New suits—Appetite for disruption in the legal world*. Stampfli Verlag.

Paraskevopoulos, D. (2021, March). *Estonian eID developers are recognized by the UN, Google; build digital societies in India and Azerbaijan*. Retrieved September 18, 2021, from https://e-estonia.com/estonian-eid-developers-are-recognized-in-india-and-azerbaijan/

Plantera, F. (2019, March). *The business registers of Estonia and Finland start cross-border interoperability*. Retrieved August 25, 2021, from https://x-road.global/case-study-the-business-registers-of-estonia-and-finland

Raconteur. (2019). *A day in data: Infographic*. Retrieved October 14, 2021, from https://www.raconteur.net/infographics/a-day-in-data/

Republic of Estonia. (2018, February). *Estonia eID scheme: ID card*. Retrieved September 18, 2021, from https://ec.europa.eu

Reynolds, M. (2016, October). *Welcome to e-Estonia, the world's most digitally advanced society*. Retrieved August 28, 2021 from https://www.wired.co.uk/article/digital-estonia

Safronov, V. (2021, June). *Mr. Freeman: The voice of freedom*. Retrieved July 30, 2021, from https://freeton.house/en/mr-freeman-the-voice-of-freedom/

Shakespeare, W. (1993) *Romeo and Juliet*. Dover Publications.

Shrier, D. (2019). Data usefulness to consumers. In T. Hardjono, D. L. Shrier, & A. Pentland (Eds.), *Trusted data: A new framework for identity and data sharing*. MIT Press.

Sporny, M., Longley, D., & Chadwick, D. (2019, November). *Verifiable credentials data model 1.0*. Retrieved September 18, 2021, from https://www.w3.org/TR/vc-data-model/#dfn-verifiable-credentials

Statista Research Department. (2021a, July). *Statista WhatsApp—Statistics & facts*. Retrieved August 28, 2021a, from https://www.statista.com/topics/2018/whatsapp/

Statista Research Department. (2021b, September). Facebook: Number of monthly active users worldwide 2008–2021b. In *Statista number of monthly active Facebook users worldwide as of 2nd quarter 2021*. Retrieved July 30, 2021, from https://www.statista.com/statistics/264810/number-of-monthly-active-facebook-users-worldwide/

The Sovrin Foundation. (2018, January). *A protocol and token for self-sovereign identity and decentralized trust*. Retrieved August 18, 2021, from https://sovrin.org/wp-content/uploads/2018/03/Sovrin-Protocol-and-Token-White-Paper.pdf

United Nations. (1948, December). Article 2. In *Universal Declaration of Human Rights*. Retrieved October 14, 2021, from https://www.un.org/en/about-us/universal-declaration-of-human-rights

Vyjayanti, T., Desai, A., & Diofasi, J. L. (2018, April). *The global identification challenge: Who are the 1 billion people without proof of identity?* Retrieved September 7, 2021, from (worldbank.org), https://www.id4africa.com/main/files/The_ID4Africa_Movement.pdf

Warwick, M. (2021, September). *Digital banking service M-Pesa is now the biggest FinTech platform in Africa.* Retrieved October 3, 2021, from https://www.telecomtv.com/content/digital-platforms-services/branchless-banking-service-m-pesa-is-now-the-biggest-fintech-platform-in-africa-42322/

Werkhäuser, N., & Esther, F. (2021). *New German ID cards: More control, less freedom?* Retrieved October 1, 2021, from https://www.dw.com/en/new-german-id-cards-more-control-less-freedom/a-58088333

Winstead, J., & Gershon, N. (2013). The world-wide web: A global source of data and information. In J. E. Dubois & N. Gersho (Eds.), *The information revolution: Impact on science and technology.* Springer.

World Bank Group. (2016). *Enabling digital development. Digital identity.* Retrieved August 28, 2021, from https://documents1.worldbank.org/curated/en/896971468194972881/310436360_20160263021000/additi onal/102725-PUB-Replacement-PUBLIC.pdf

World Legal Summit. (2019). *Decentralized identity readiness index.* Retrieved August 2021, from https://worldlegalsummit.org/decentralized-identity-rea diness-index

Zyskind, G., Nathan, O., & Pentland, A. (2021). *Enigma: Decentralized computation platform with guaranteed privacy.* Retrieved August 18, 2021 from https://web.media.mit.edu/~guyzys/data/enigma_full.pdf

Proposals for a New Legal Framework on AI: Possible Discrepancies Between Liability, Compliance and Duties of Company Directors

Francesco Pacileo

Abstract Two recent proposals for an EU Regulation on AI, both of which adopt a risk-based approach, should be aligned with one another. However, the no-fault liability regime concerning high-risk AI systems appears to be disconnected from the duties of compliance inspired by the Ethical Guidelines for Trustworthy AI. Furthermore, the duties of company directors, and in particular the duty to act on an informed basis and the duty to overview when using non-high-risk AI systems for important decision-making, could be a concrete obstacle to the complete success of these proposals. In this regard, a human-centric approach should help to solve these issues.

F. Pacileo (✉)
Sapienza University of Rome, Rome, Italy
e-mail: francesco.pacileo@uniroma1.it

M. Bertolaso et al. (eds.), *Digital Humanism*,
https://doi.org/10.1007/978-3-030-97054-3_10

Keywords Artificial intelligence · No-fault liability · Company · Directors' duties · Duty to act on an informed basis · Decision-making · Risk-based approach · Innovation

1   Introduction

This chapter aims to share some first impressions concerning the recent Proposal for a Regulation on AI, provided by the European Commission in April 2021, and possible discrepancies between this and another recent Proposal for a Scheme of Regulation on liability when deploying AI systems, which was provided by European Parliament in October 2020. It also seeks to highlight possible discrepancies between the AI Act and the duties of company directors when deploying AI systems for decision-making.

The European Commission has recently published a Proposal for a Regulation of the European Parliament and of the Council which sets out harmonized rules on artificial intelligence (the AI Act) and amends specific EU legislative acts (European Commission, 2021).

At first glance, the proposal concerns, inter alia, new rules to address the specific risks posed by AI systems and sets standards for developing a so-called Trustworthy AI. In particular, the AI Act foresees many mandatory duties for the deployers of certain AI systems.

The other proposal, for a Scheme of Regulation of the European Parliament and of the Council on liability for the operation of artificial intelligence systems (European Parliament, 2020) foresees specific liability regimes when deploying AI systems.

Both proposals are part of the wider European digital strategy, which aims to put digital tools to work in delivering public goods to European citizens (European Commission, 2020d; also, 2018, 2019a, 2019b, 2020e; European Parliament, 2017). This is a further step towards a European "digital constitutionalism", composed also of the GDPR and, in time, the proposed Digital Services Act and the proposed Digital Markets Act (European Commission, 2020a, 2020b)—which, once adopted, will regulate online platforms and services—and of the proposed Data Governance Act (European Commission, 2020c; also Celeste, 2019; De Gregorio, 2021; Floridi, 2021).

This chapter uses the term "deploy/-er/-ment" to refer to any "provider", "programmer", "manufacturer", "operator", "user" or "vendor", and to all the subjects (and correlative verbs) who can have any control over the risks related to AI systems.[1]

The chapter is divided into three parts. First, certain aspects of the working AI Act are briefly analyzed to highlight many possible critical issues in its own regulation. Second, possible discrepancies between the two proposals mentioned above are examined; and finally, possible discrepancies between the AI Act and the duties of company directors when deploying AI systems for decision-making are analyzed.

The following analysis aims to help to outline the correct approach to setting up a legal framework on AI. In this regard, a human-centric approach that gives central relief to human rights could be of fundamental importance.

2   A Risk-Based Approach

It is important to note that both the proposals pursue a risk-based approach (RBA) that differentiates "high-risk" AI systems from "non-high-risk"/ "limited-risk"/ "minimum-risk" (hereafter "non-high-risk") AI systems.

This RBA is promoted by the European Commission and by the High-Level Expert Group on AI (HLEG) (European Commission, 2020e; HLEG, 2019b, 37 et seq.).

Nonetheless, an examination of both proposals shows that they use different criteria to distinguish high-risk and non-high-risk AI systems (also German Data Ethics Commission, 2019, 173 et seq.).

In fact, Art. 3 of the Scheme of Regulation on liability defines "high-risk" as follows:

> [A] significant potential in an autonomously operating AI-system to cause harm or damage to one or more persons in a manner that is random and goes beyond what can reasonably be expected; the significance of the potential depends on the interplay between the severity of possible harm or damage, the degree of autonomy of decision-making, the likelihood

---

[1] See Arts. 2 and 3 AI Act (European Commission, 2021). See also the definitions of "operator" provided by Art. 3 Scheme of Regulation on liability (European Parliament, 2020).

that the risk materializes and the manner and the context in which the AI-system is being used.

Furthermore, the AI Act, which also distinguishes a limited number of "unacceptable-risk" AI systems which are banned, includes the high-risk AI systems in a close taxonomic list. High-risk AI systems concern AI technology that can have a significant impact on health, safety or fundamental human rights of natural persons,[2] which includes:

– Remote biometric identification systems.
– Critical infrastructure (e.g., transport), that could put the life and health of citizens at risk
– Educational or vocational training that may determine access to education and professional course of someone's life (e.g., exam marks)
– Safety components of products (e.g., AI application in robot-assisted surgery)
– Employment, workers management and access to self-employment (e.g., CV-sorting software for recruitment procedures)
– Essential private and public services (e.g., credit scoring denying citizens the opportunity to obtain a loan)
– Law enforcement that may interfere with people's fundamental rights (e.g., evaluation of the reliability of evidence)
– Migration, asylum, and border control management (e.g., verification of the authenticity of travel documents)
– Administration of justice and democratic processes (e.g., applying the law to a concrete set of facts).

3   The AI Act and the Duty of Compliance Only for High-Risk AI Systems

On the pathway of the described RBA, the AI Act foresees, among others, a mandatory duty of compliance and an *ex-ante* conformity assessment *only for high-risk AI systems*, while for the non-high-risk AI systems

---

[2] And that can be a safety component of a product or itself a product (i) covered by the EU safety legislation and required to undergo third-party conformity assessment with a view to the placing on the market or putting into service of that product pursuant to EU legislation, or (ii) included in a further taxonomic list. See Art. 6 AI Act.

it provides (together with a slight duty of transparency for those that interact with natural persons) that the deployers can adopt a purely voluntary code of conduct.

Moreover, such duty of compliance requires the adoption of adequate risk assessment and mitigation systems, and high-quality datasets, but above all, it is widely inspired by and contained in the Ethics Guidelines for Trustworthy AI provided by the HLEG which were set up by the European Commission in 2019 (HLEG, 2019a; also International Technology Law Association, 2019; Mökander & Floridi, 2021).

As a matter of fact, the Ethics Guidelines pursue a *human-centric approach*, giving central relief to *human dignity*, intended also to include the capability of self-assessment.[3] In particular, the Guidelines foresee many requirements for AI systems, among which are respect for human autonomy, prevention of harm, fairness, transparency, explicability, human agency and oversight, technical robustness and safety, privacy and data governance, accountability, and social and environmental well-being.[4] Following the same approach, the European Commission has explicitly declared that "AI is not an end in itself, but a *tool that has to serve people* with the ultimate aim of increasing human well-being" (European Commission, 2019a, 1).[5]

In spite of being in agreement with such an approach for high-risk AI systems (Floridi, 2021),[6] exists the risk that the AI Act could imply *substantial deregulation for all of the other non-high-risk AI systems*, whose specific provisions refer to the soft law or sectoral regulation.

Due to the close taxonomic list including every high-risk AI system, a huge number of AI systems could be classed as non-high-risk.

---

[3] See Recitals 17, 28, 37 AI Act and Para. 3.5. Explanatory Memorandum of AI Act, annexed to European Commission (2021).

[4] See also Recitals 47 et seq. AI Act and Para. 1.1. Explanatory Memorandum of AI Act, annexed to European Commission (2021).

[5] Emphasis added. For a different approach, see Teubner (2018).

[6] Yet he is critical of the expression "human-centric" which, in his opinion, seems synonymous with "anthropocentric", meaning centrality of humanity in the environment as if everything must always be at its service, including every aspect of the natural world, no matter the cost. Nonetheless, "human-centric" in the AI legal framework context has a clearly different meaning, so that this author appears to be too politically correct.

A Good End with Bad Means?

The *ratio legis* pursued by the AI Act has as its goal, among others, to promote a single market on AI, to foster innovation, preventing "chilling effects" (i.e., a restraint to innovation) due to an excess of regulation and to legal uncertainty, and to boost EU competitiveness.[7]

In pursuance of these aims, the AI Act seeks to deploy proportionate and flexible rules and therefore foresees a mandatory duty for high-risk AI systems, yet substantial deregulation for non-high-risk AI systems.

As far as EU competitiveness is concerned, scholars and experts are certainly aware of the ongoing global competition in digital innovation. The main competitors of the EU are China, the US, and private big-tech companies. In such a contest, the EU could play a fundamental role as supervisor of fundamental human rights. Furthermore, the EU legal framework on AI could also have an impact on the deployers of AI systems outside the EU, whereby they comply with EU regulations even in other countries because it is more practical to have a single approach globally and it would be more difficult to explain the adoption of less qualitative standards in other markets (the so-called "Bruxelles effect"). In other words, the "Bruxelles effect" enables the EU to extend de facto its legal framework outside its boundaries through market mechanisms (Bradford, 2020).

As a result, is the EU pursuing a good end with bad means?

4   Possible Discrepancies Between the AI Act and the Proposed Regulation on Liability for Deploying AI Systems

This last question brings us to the second issue considered in this chapter: possible discrepancies between the AI Act and the proposal of the regulation of liability regimes for AI.

The latter proposal, still pursuing an RBA, foresees *a no-fault liability regime for high-risk AI systems* but a fault liability for non-high-risk AI systems, although with the reversal of the burden of proof on the part of the deployer.

---

[7] See Recitals 1, 2, 5 et seq. AI Acts and Paras. 1.1., 1.3., 2.1., 2.3. Explanatory Memorandum of AI Act, annexed to European Commission (2021).

What, then, should the deployers of a high-risk AI system do to prevent no-fault liability for damages caused by the system, even though they have complied with the terms of the AI Act?

At first glance, it appears that the duty of compliance (foreseen by the AI Act) and the no-fault liability regime should be coordinated. The AI Act proposed by the Commission should be set up (also) by the European Parliament, which has itself provided the Scheme of Regulation on liability when deploying AI systems. It is difficult to imagine that the Parliament would not emend the formulation of this proposal in view of its apparent inconsistency with the Scheme.

5   Possible Discrepancies Between the AI Act and the Duties of Company Directors When Deploying AI Systems for Decision-Making

Finally, the possible deregulation of non-high-risk AI systems could collide, among others, with the duty of company directors when deploying AI for crucial decision-making.

Indeed, according to company law in the most developed jurisdictions, company directors must also comply with (inter alia) the duty of care when deploying AI systems. In turn, such duty implies the duty of due diligence and the *duty to act on an informed basis.*[8]

---

[8] In the US, see the famous case law *Smith v. Van Gorkom*, 488 A.2d. 858 (Del. 1985), in which the directors were held responsible for not having assumed adequate information concerning a merger. See Möslein (2018, 661). See also *In re Caremark International Inc.*, Derivative Litigation, 698 A.2d 959 (Del. Ch., 1996) ["information and reporting systems ... that are reasonably designed to provide to senior management and to the board itself timely, accurate information sufficient to allow management and the board, each within its scope, to reach informed judgments concerning both the corporation's compliance with law and its business performance"]. See, in this regard, Armour and Eidenmüller (2020, 103 ss.), and the literacy mentioned therein. Also in France, scholars and judges consider the duty to act on an informed basis. See Le Cannu and Dondero (2015, 478), and (concerning the responsibility of directors in general) 310 et seq., 495 et seq. (concerning Art. L. 225–35 al. 3 *c.comm.* and the duty of the *président* or of the *directeur général* of a *SA* to keep the directors informed). In Germany, § 93 *Abs.* 1 *Satz* 2 *AktG*, entitled 'Sorgfaltspflicht und Verantwortlichkeit der Vorstandsmitglieder', foresees that "*Eine Pflichtverletzung liegt nicht vor, wenn das Vorstandsmitglied bei einer unternehmerischen Entscheidung* vernünftigerweise *annehmen durfte, auf der Grundlage angemessener Information zum Wohle der Gesellschaft zu handeln*" (emphasis added). See Möslein (2018, 662 also *sub* fn. 66). In Spain, Art. 226 LSC, foresees that "*En el ámbito de las decisiones estratégicas y de negocio, sujetas a la discrecionalidad empresarial, el*

Moreover, some scholars provide a wide interpretation of case law and the provisions made by principal jurisdictions concerning whether directors can delegate tasks to employees or third parties, by assuming that they can *delegate* decision rights *to AI systems* (Abriani, 2020, 270 et seq.; Möslein, 2018, 658 et seq.). Always assuming that such interpretation is correct, on the one hand directors cannot delegate the management function itself and, on the other hand, they retain the *duty of oversight*.[9] In other words, the core management decisions must always remain within the board of directors, which still has a duty to supervise completion of the delegated tasks.

Are these duties consistent with deregulation on non-high-risk AI systems?

The following example illustrates this possible discrepancy. A listed public company, with thousands of employees and thousands of small investors, is in crisis and its directors deploy a non-high-risk AI system (non-high-risk, just because it is not included in the taxonomic list). The system has a high degree of accuracy but also a black-box algorithm (Bathaee, 2018; Burrell, 2016, 4 et seq.; European Commission, 2020e, 11; Hildebrandt, 2016, 26; Karnow, 1996, 161, 176 et seq.; Kroll et al., 2017; Pasquale, 2016). This AI system can provide very significant insights about the evolution of the market of interest and about many important factors strictly related to the company. In other words, the black-box non-high-risk AI system is fundamental to the directors' decision whether or not to wind up the company.

---

estándar de diligencia de un ordenado empresario se entenderá cumplido cuando el administrador haya actuado de buena fe, sin interés personal en el asunto objeto de decisión, con información suficiente y con arreglo a un procedimiento de decisión adecuado" (emphasis added). In Italy, Art. 2381 c.c. foresees explicitly the duty of directors to act on an informed basis. See Angelici (2006; 2012, 401 et seq.).

[9] See Delaware General Corporation Law § 141(a) ["The business and affairs of every corporation organized under this chapter shall be managed by *or under the direction of* a board of directors..." (emphasis added)]; *In re Caremark* (fn. 16); *Marchand v. Barnhill* (Del. 2019) 2019 WL 2509617. In this regard, see Armour and Eidenmüller (2020, 103 ss.) and Möslein (2018, 661–662). In France, legal scholars point out that directors cannot escape from liability for managing the business by delegating their tasks. Nonetheless, according to the case law, directors may prevent liability for harms caused by agents if they have, cumulatively: (i) assessed that competencies, powers and legal tool of the agents are suitable; (ii) overseen the selected agents; (iii) adopted the necessary provisions, when they become aware of reprehensible behaviour on the part of the agents. See Le Cannu and Dondero (2015, 316 and 336).

In this regard, several legal scholars think that, when deploying AI systems for crucial decision-making, directors should be (also indirectly) tech-savvy or at least tech-friendly (Armour & Eidenmüller, 2020, 101 et seq.; Möslein, 2018, 660). That means they should know and understand the input data, the logic of the output or even the source code of the algorithm (Abriani, 2020, 272–273; Montagnani, 2020, 80; Mosco, 2019, 255). That implies a certain degree of transparency and requirements not far from the concepts of fairness, explicability, robustness, safety, etc., foreseen by the Ethics Guidelines on Trustworthy AI (and consequently by the proposed AI Act, even if only for high-risk AI systems) (Mökander & Floridi, 2021).

This is a very complex issue, the resolution of which requires multidisciplinary competencies. Consider, for example, the difficulties of applying the EU Directive 2016/943 on the protection of undisclosed know-how and business information against their unlawful acquisition, use and disclosure, to algorithms protected by trade secrets (Diakopoulos, 2014, 26–27; Montagnani, 2020, 79–80),[10] and the GDPR specifically to automated processing data (Kroll et al., 2017, 691).[11]

Some scholars who study the GDPR and the right of the data subject to contest decisions based solely on automated processing, adopt the term "legibility" of the algorithm, citing "the capability of individuals to autonomously understand data and analytics algorithms, with a concrete comprehension of methods and data used" (Malgieri & Comandè, 2017, 245, 256 et seq.; Mortier et al., 2015).[12]

Also in connection with the GDPR, other scholars elaborate the concept of "counterfactual explanation" with reference to a diverse set of different choices of nearby possible worlds for which the counterfactual holds or a preferred outcome is delivered (Watcher et al., 2018).[13]

---

[10] The latter author refers to the US Digital Millennium Copyright Act (DMCA), as well as to the praxis of software vendors arranging anti-reverse engineering clauses for final users.

[11] According to which "privacy may be at risk from an automated decision that reveals sensitive information just like fairness may be at risk from an automated decision".

[12] For a different interpretation of the GDPR, see Watcher et al. (2017).

[13] An example of counterfactual statement could be: "You were denied a loan because your annual income was € 30,000. If your income had been €45,000, you would have been offered a loan."

Such "legibility" and/or "counterfactual explanation" should also be appropriate for the duty of directors to act on an informed basis and the duty of oversight when using AI systems.

To sum up, this implies a sort of mandatory duty of directors to comply with (at least) technical standards for a Trustworthy AI foreseen by the Ethics Guidelines. According to Italian company law, this duty should be included in the principles of sound management of a business foreseen by Art. 2497 c.c.[14]

Furthermore, the *human-centric approach* and its central relief to human rights given by the Ethics Guidelines arrive at the same conclusions. In particular, in considering AI purely as a *tool to serve people*, human dignity, intended as the capability of self-assessment, as well as the respect for human autonomy, provides that directors must have—and explain, at least within the board—a *personal and critical* (no matter whether positive or negative) *opinion* about the output of an AI system. An *ipse dixit* approach is not permitted. Indeed, the ability to determine and balance goals and values (such as shareholder value, stakeholder interests—especially those of the creditors, employees and investors—and even social and environmental sustainability),[15] together with the emotional element, are typical of and exclusive to humans. Hence, they cannot

---

[14] See also in Germany § 93 *AktG* ("*Die Vorstandsmitglieder haben bei ihrer Geschäftsführung* die Sorgfalt eines ordentlichen und gewissenhaften Geschäftsleiters *anzuwenden*") and in the UK Sec. 174 Companies Act 2006 ("care, skill and diligence that would be exercised by a *reasonably diligent person* with … the general knowledge, skill and experience *that may reasonably be expected of a person carrying out the functions* carried out by the director in relation to the company") (all emphasis added).

[15] In this regard, see again § 93 *AktG*, according to which the directors adopt a business judgement "*zum Wohle der Gesellschaft*". See also Sec. 172 UK Companies Act 2006 entitled 'Duty to promote the success of the company' ["A director of a company must act in the way he considers, *in good faith*, would be most likely to promote the *success of the company for the benefit of its members as a whole*, and in doing so have regard (among other matters) to (a) the likely consequences of any decision in the long term, (b) the interests of the company's employees, (c) the need to foster the company's business relationships with suppliers, customers and others, (d) the impact of the company's operations on the community and the environment, (e) the desirability of the company maintaining a reputation for high standards of business conduct, and (f) the need to act fairly as between members of the company"] (all emphasis added). See finally Art. 1833 of the French Code Civil, recently amended ["*La société est gérée dans son* intérêt social, *en prenant en considération les enjeux* sociaux *et* environnementaux *de son activité*" (emphasis added)] and summarized in Art. L225-35 and L225-64 of the Code de Commerce, related to the duties of the *conseil d'administration* and the *directoire* of the *sociétés anonymes*.

be substituted by an algorithm (Armour & Eidenmüller, 2020, 101; Burbidge et al., 2020, 154 et seq.; Kemper & Kolkman, 2019; Möslein, 2018, 660).

6 Conclusion

That brings us to the final (first) conclusion, which will be introduced with a further question. Is company law a limitation on the New Regulatory Framework on AI, when company law provides mandatory duties also for non-high-risk AI systems?

From an empirical point of view, most programmers, providers, manufacturers, operators, users, vendors, and all the components of the AI value chain, are companies. Nevertheless, company law (together with the GDPR) should be regarded as a good incentive to improve the New Regulatory Framework on AI, always on the pathway of a human-centric approach (Möslein, 2018, 650 et seq.).

Of course, this is a matter for further complex analysis, but a possible normative foothold for reconciling company law with the proposed AI Act could be a wide interpretation of the adverse impact that a high-risk AI system would have on "fundamental human rights", foreseen by Art. 7.1.(b).

References

Abriani, N. (2020). La *corporate governance* nell'era dell'algoritmo. Prolegomeni a uno studio sull'impatto dell'intelligenza artificiale sulla *corporate governance*. *Il Nuovo Diritto delle Società*, *18*(3), 261–286.

Angelici, C. (2006). Diligentia quam in suis e business judgement rule. *Rivista del diritto commerciale*, *104*(I), 675–693.

Angelici, C. (2012). *La società per azioni. Principi e problemi*, vol. I. In *Trattato di diritto civile e commerciale*, a cura di P. Schlesinger, Giuffrè.

Armour, J., & Eidenmüller, H. (2020). Self-driving corporations? *Harvard Business Law Review*, *10*, 87–110.

Bathaee, Y. (2018). The artificial intelligence black box and the failure of intent and causation. *Harvard Journal of Law & Technology*, *31*(2), 889–938.

Bradford, A. (2020). *The Bruxelles effect*. Oxford University Press.

Burbidge, D., Briggs, A., & Reiss, M. J. (2020). *Citizenship in a networked age: An agenda for rebuilding our civic ideas*. University of Oxford Research Project. Report retrieved November 1, 2021, from https://citizenshipinan etworkedage.org

Burrell, J. (2016, January–June). How the machine 'thinks': Understanding opacity in machine learning algorithms. *Big Data & Society*, 1–12. https://doi.org/10.1177/2053951715622512

Celeste, E. (2019). Digital constitutionalism: A new systematic theorisation. *International Review of Law, Computers & Technology, 33*, 76–99. https://doi.org/10.1080/13600869.2019.1562604

De Gregorio, G. (2021). The rise of digital constitutionalism in the European Union. *International Journal of Constitutional Law., 19*(1), 41–70. https://doi.org/10.1093/icon/moab001

Diakopoulos, N. (2014). *Algorithmic accountability reporting: On the investigation of black boxes*. Tow Center for Digital Journalism Publications, Columbia University. https://doi.org/10.7916/D8ZK5TW2

European Commission. (2018). *Artificial intelligence for Europe*. Brussels, 25.4.2018. COM(2018) 237 final. Retrieved November 1, 2021 from https://eur-lex.europa.eu/legal-content/EN/TXT/PDF/?uri=CELEX:520 18DC0237&from=EN

European Commission. (2019a). *Building trust in human-centric artificial intelligence*. COM(2019a) 168 final, Brussels, 8.4.2019a. Retrieved November 1, 2021 from https://eur-lex.europa.eu/legal-content/EN/TXT/PDF/?uri=CELEX:52019DC0168&from=IT

European Commission. (2019b). *Report on the safety and liability implications of artificial intelligence, the Internet of things and robotics*. Brussels, 19.2.2020 COM(2020) 64 final. Retrieved November 1, 2021 from https://eur-lex.europa.eu/legal-content/EN/TXT/PDF/?uri=CELEX:52020DC0064&from=IT

European Commission. (2020a). *Proposal for a Regulation of the European Parliament and of the Council on a single market for digital services (Digital Services Act) and amending Directive 2000/31/EC*. Brussels, 15.12.2020a COM(2020a) 825 final. Retrieved November 1, 2021 from https://eur-lex.europa.eu/legal-content/EN/TXT/PDF/?uri=CELEX:52020PC0825&from=en

European Commission. (2020b). *Proposal for a Regulation of the European Parliament and of the Council on contestable and fair markets in the digital sector (Digital Markets Act)*. Brussels, 15.12.2020b COM(2020b) 842 final. Retrieved November 1, 2021 from https://eur-lex.europa.eu/legal-content/EN/TXT/PDF/?uri=CELEX:52020PC0842&from=en

European Commission. (2020c). *Proposal for a Regulation of the European Parliament and of the Council on European data governance (Data Governance Act)*. Brussels, 25.11.2020c COM(2020c) 767 final. Retrieved November 1, 2021 from https://eur-lex.europa.eu/legal-content/EN/TXT/PDF/?uri=CELEX:52020PC0767&from=EN

European Commission. (2020d). *Shaping Europe's digital future*. Luxembourg, 19.2.2020d. Retrieved November 1, 2021 from https://ec.europa.eu/info/publications/communication-shaping-europes-digital-future_it

European Commission. (2020e). *White paper on artificial intelligence—A European approach to excellence and trust*. Brussels, 19.2.2020e. COM(2020e) 65 final. Retrieved November 1, 2021 from https://eur-lex.europa.eu/legal-content/EN/TXT/PDF/?uri=CELEX:52020DC0065&from=EN

European Commission. (2021). *Proposal for a Regulation of the European Parliament and of the Council laying down harmonised rules on artificial intelligence (the AI Act) and amending certain Union legislative acts*. Brussels, 21.4.2021. COM(2021) 206 final. Retrieved November 1, 2021 from https://eur-lex.europa.eu/resource.html?uri=cellar:e0649735-a372-11eb-9585-01aa75ed71a1.0001.02/DOC_1&format=PDF and https://eur-lex.europa.eu/resource.html?uri=cellar:e0649735-a372-11eb-9585-01aa75ed71a1.0001.02/DOC_2&format=PDF

European Parliament. (2017). *Civil Law Rules on robotics. Resolution of 16 February 2017 with recommendations to the Commission on Civil Law Rules on Robotics (2015/2103(INL)*. P8_TA(2017)0051. Retrieved November 1, 2021 from https://eur-lex.europa.eu/legal-content/EN/TXT/PDF/?uri=CELEX:52017IP0051&from=IT

European Parliament. (2020). *Civil liability regime for artificial intelligence*. Resolution of 20.10.2020 with recommendations to the Commission on a civil liability regime for artificial intelligence (2020/2014(INL)). P9_TA(2020)0276. Retrieved November 1, 2021 from https://www.europarl.europa.eu/doceo/document/TA-9-2020-0276_EN.pdf

Floridi, L. (2021). The European legislation on AI: A brief analysis of its philosophical approach. *Philosophy & Technology, 34*, 215–222. https://doi.org/10.1007/s13347-021-00460-9

German Data Ethics Commission. (2019). *Opinion of the Data Ethics Commission*. Retrieved November 1, 2021 from https://www.bmjv.de/SharedDocs/Downloads/DE/Themen/Fokusthemen/Gutachten_DEK_EN_lang.pdf?__blob=publicationFile&v=3

Hildebrandt, M. (2016). Law as information in the era of data-driven agency. *The Modern Law Review, 79*(1), 1–30.

HLEG. (2019a). *Ethics guidelines for trustworthy AI*. Brussels, 8.4.2019a. Retrieved November 1, 2021 from https://digital-strategy.ec.europa.eu/en/library/ethics-guidelines-trustworthy-ai

HLEG. (2019b). *Policy and investment recommendations for trustworthy AI, 2019b*. Brussels, 26.6.2019b. Retrieved November 1, 2021 from https://digital-strategy.ec.europa.eu/en/library/policy-and-investment-recommendations-trustworthy-artificial-intelligence

International Technology Law Association. (2019). *Responsible AI*, C. Morgan (ed.), McLean Virginia. Retrieved November 1, 2021 from https://www.ite chlaw.org/sites/default/files/Responsible_AI.pdf

Karnow, C. E. A. (1996). Liability for distributed artificial intelligences. *Berkeley Technology Law Journal, 11*(1), 147–204.

Kemper, J., & Kolkman, D. (2019). Transparent to whom? No algorithmic accountability without a critical audience. *Information, Communication & Society, 22*(14), 2081–2096. https://doi.org/10.1080/1369118X.2018.147 7967

Kroll, J. A., Huey, J., Barocas, S., Felten, E. W., Reidenberg, J. R., Robinson, D. G., & Yut, H. (2017). Accountable algorithms. *University of Pennsylvania Law Review, 165*(3), 633–706.

Le Cannu, P., & Dondero, B. (2015), *Droit des sociétés* (6th ed.). L.G.D.J.

Malgieri, G., & Comandè, G. (2017). Why a right to legibility of automated decision-making exists in the general data protection regulation. *International Data Privacy Law, 7*(4), 243–265. https://doi.org/10.1093/idpl/ipx019

Mökander, J., & Floridi, L. (2021). Ethics-based auditing to develop trustworthy AI. *Minds and Machines., 31*, 323–327. https://doi.org/10.1007/s11023-021-09557-8

Montagnani, M. L. (2020). Flussi informativi e doveri degli amministratori di società per azioni ai tempi dell'intelligenza artificiale. *Persona e Mercato, 2*(2), 31/66–84.

Mortier, R., Haddadi, H., Henderson, T., McAuley, D., & Crowcroft, J. (2015). *Human data interaction: The human face of the data-driven society*. Retrieved November 1, 2021 from arXiv:1412.6159v2 and from https://ssrn.com/abs tract=2508051

Mosco, G. D. (2019). Roboboard. L'intelligenza artificiale nei consigli di amministrazione. *Analisi Giuridica dell'Economia, 18*(1), 247–260. https://doi.org/10.1433/94555

Möslein, F. (2018), Robots in the boardroom: Artificial intelligence and corporate law. In W. Barfield & U. Pagallo (Eds.), *Research handbook on the law of artificial intelligence* (pp. 649–670). Edward Elgar. https://doi.org/10.4337/9781786439055.00039

Pasquale, F. (2016). *The black box society*. Harvard University Press.

Teubner, G. (2018). Digitale Rechtssubjekte? Zum privatrechtlichen Status autonomer Softwareagenten. *Archiv für die civilistische Praxis, 218*, 155–205. https://doi.org/10.1628/acp-2018-0009

Watcher, S., Mittelstadt, B., & Floridi, L. (2017). Why a right to explanation of automated decision-making does not exist in the general data protection regulation. *International Data Privacy Law, 7*(2), 76–99. https://doi.org/10.1093/idpl/ipx005

Watcher, S., Mittelstadt, B., & Russell, C. (2018). Counterfactual explanations without opening the black box: Automated decisions and the GDPR. *Harvard Journal of Law & Technology, 31*(2), 841–887.

PART IV

Individual and Social Impacts of Digital Technologies

CHAPTER 11

Education and Artificial Intelligence: A New Perspective

Pierluigi Malavasi and Cristian Righettini

Abstract Education and artificial intelligence? The question forces us to ask ourselves about the limits and potential of machines and algorithms, or rather about the responsibilities of our choices, about their effects. Technological devices that are increasingly sophisticated, autonomous and integrable with our organism are simulating human capabilities. What we call artificial intelligence marks everyday life and constitutes a challenge for the future of civilization. Challenge and therefore unknowns, risks and opportunities. A new alliance between education and artificial intelligence means cultivating people's creative and civic resources in relational contexts where digital connectivity is so pervasive that it is unthinkable to interpret its logic without adequate pedagogical awareness. It is a question

---

This chapter was conceived in a unitary way; the first and second sections were drafted by Pierluigi Malavasi and the third by Cristian Righettini.

P. Malavasi (✉) · C. Righettini
Catholic University of the Sacred Heart, Milan, Italy
e-mail: pierluigi.malavasi@unicatt.it

 M. Bertolaso et al. (eds.), *Digital Humanism*,
https://doi.org/10.1007/978-3-030-97054-3_11

of educating to discernment and understanding how radical innovations, of process and product, can contribute to the common good in addressing educational needs and social fragility, inequality and poverty. A pedagogy of artificial intelligence urges us to take care of the human. It speaks to parents, teachers and those involved in design, manufacture and training of intelligent machines, encouraging them to foster the development of a responsible conscience that will give rise to fair values and actions.

Keyword Education · Artificial intelligence · Robotics · New humanism · Pedagogy

1   Ambiguity and Power of Technological Civilization: Education and Artificial Intelligence, Representative Issues

"We are both intrigued and frightened by the prospect of machines that can respond to us as a person would and, on a certain level, might even seem human" (Brooks, 2017, p. 52, trans. by the authors). It is within the framework of the history of human development that this volume aims to situate the relationship between civilization and instrumental devices, between the mystery of the person and the efficiency of machines.

Robotics was created in the middle of the last century as a sub-sector of cybernetics: it has become firmly integrated with engineering research and established itself through the design of machines equipped with increasingly autonomous systems of information, assessment and action. The growing media interest in robots, often marked by concerns and worries, is matched by their widespread use in many industrial fields. The profound anthropological changes of the contemporary world, such as genetics and neuroscience research, raise unprecedented questions (Ravasi, 2017). However, Rodney Brooks, long-time director of the Massachusetts Institute of Technology's Computer Science and Artificial Intelligence Laboratory, has this to say (and we share his opinion):

> Is artificial intelligence destined to dominate our lives? There's a lot of hysteria in the debate about the future. Many people wonder how powerful robots will become, when it will happen, and what will happen to our jobs

.... Almost every innovation in the field of robotics and artificial intelligence takes much longer to spread than predicted by the experts and external observers. ... We won't be caught by surprise, I'm not saying there will never be problems, I'm saying they won't be sudden and unexpected. (Brooks, 2017, pp. 59–61, trans. by the authors)

The most popular representations of so-called intelligent systems include the influential mythology of metallic automatons taking over from humankind, or robots replacing human work activities and developing malevolent or violent attitudes. These stereotypes have been nurtured since the 1950s by science fiction, literature and filmography, which expressed a reaction to the astonishing scientific and technological advances and the continuing risk of nuclear catastrophe.

Open thinking and multilateral reflection on innovation have to face major challenges, while avoiding carelessness and superficiality. Several issues are at stake. Among others, it is worth mentioning the very ambiguity of words such as "artificial intelligence" (Accoto, 2017) and "robotics"; the media overestimate of the short-term effects of technologies; the real skills of robots and the likely improvement of their performance; the expectation/concern about the timing of deployment and employment spin-offs; and the popularity of science fiction mythology and, at the same time, the need for technological literacy/education.

The expression "educated robot" concerns industrial and service organization in the contemporary world and alludes to the need to include, among the purposes of human education, also the activities aimed at the conception, production and operational management of machines, supports and applications. First of all, it is a matter of becoming aware of what Adam Greenfield (2017) defines as "radical technologies" or the design of everyday life in contact with tools equipped with advanced functions and, paradoxically, "intelligent". The smartphone – which has become almost indispensable to organize our daily existence – is, in a protean way, the symbol *par excellence* of the modern human relationship with a machine.

Very few objects have been so omnipresent and pervasive in the history of civilizations (Ferraris, 2005). For many of us, the smartphone is the last thing we pay attention to before we fall asleep and the first thing we pick up when we wake up. We use it to meet people, to communicate, to have fun, to orient ourselves, to buy and to sell. However, we also use it to document the places we go, the things we do, our acquaintances; to fill the empty spaces, the moments of pause and the silences that used to occupy part of our existence.

> Smartphones have effectively altered the entire fabric of daily life, by completely reorganizing longstanding spaces and rituals, and transforming others beyond recognition. At this historical juncture, it is simply impossible to understand the ways through which we know and practice the world around us without having at least some idea of how the smartphone works and the various infrastructures on which it depends. Its ubiquity, nevertheless, makes it an object that is anything but trivial. We use it so often that we fail to clearly understand what it really is; it has appeared in our lives so suddenly, and in such an all-encompassing way, that the scale and force of the changes it has brought about largely elude our awareness. (Greenfield, 2017, p. 7)

Although, strictly defined, a smartphone is not a robot, it is difficult to deny that it represents a formidable emblem of the performative relevance of technology. Using networked digital information, it has become the dominant way in which we experience everyday life. In educating people to use smartphones responsibly, a window of sense is opened on the web of technical, financial, legal and operational connections and agreements that constitute not only a technological device but an ecosystem of services. Equally important are pedagogical analysis and ethical-educational discernment regarding the complexity of issues posed by the planetary network of perceptions and responses that we call the Internet of Things: here, computability and data communication are embedded and distributed within our environment, in its entirety (Kuniavsky, 2010).

The Internet of Things, in many respects analogous to the smartphone, is an assemblage of technologies, perception regimes and operating protocols identified with an all-encompassing expression familiar to a large part of public opinion. What unites very heterogeneous elements is a conception of the world that connects devices, applications, supplier companies, performance of products and services, in order to connect everyday life situations to the network, and make them available for analysis and processing.

> Although this colonization of the quotidian may be perceived as something that develops autonomously, without manifest guidance or other urgent justification than the fact that it is our technology that makes it possible,

it is always better to keep in mind how certain ambitions come into play. Some of these relate to commercial needs .... Others are based on a set of interests that may have to do with the management of the infrastructure that secures public utility goods. Inevitably, some of these ambitions involve surveillance, security, and control. (Greenfield, 2017, pp. 32–33)

We are heirs to two centuries of progress that express the extraordinary creativity of humankind. Nanotechnology, robotics and biotechnology make the availability of products and services to improve the quality of life possible. However, it cannot be ignored that never before have knowledge and especially economic power offered such effective tools for the manipulation of consciences and the domination of the whole world. There is a tendency to believe that any acquisition of power is simply an increase in well-being and life force. Our age tends to develop poor self-awareness of its own limitations: "the possibility of misusing power is constantly increasing when there are no norms of freedom, but only claims to necessity, utility and security" (Guardini, 1965, p. 87, trans. by the authors). And there is nothing to ensure that humanity will make fair and supportive use of the digital revolution, big data, and robotics in the future (Valentini et al., 2013).

> Our immense technological development has not been accompanied by a development in human responsibility, values and conscience ... We need but think of the nuclear bombs dropped in the middle of the twentieth century, or the array of technology which Nazism, Communism and other totalitarian regimes have employed to kill millions of people, to say nothing of the increasingly deadly arsenal of weapons available for modern warfare. In whose hands does all this power lie, or will it eventually end up? It is extremely risky for a small part of humanity to have it. (Francesco, 2015, pp. 104–105)

The need for an alliance between robotics and pedagogy is in the facts: a historical analysis of the use of technology shows the necessity of a constant relationship between the discourse on education and the development of technology, between training on how to use tools and ethical-moral consciousness. The close relationship with machines endowed with advanced functions in the concreteness of daily life implies a new and deep awareness of the human potential that has given rise to intelligent devices—such as the smartphone—and to real robots. The responsibility for accessing and using sensitive information, for the most diverse uses, now affects anyone using the Internet.

The notions of algorithm and computational thinking are more and more widespread in the discourse on the frontiers of education. Among the main reasons crossing the pages of this volume there is a recognition of the widening of the educational sphere that the pressing digitalization of everyday life entails. In a peculiar way, the task of thinking about the relationship between human and machine learning, before assuming a didactic and operational relevance, has a fundamental importance in the field of pedagogical reflection. The relationship, the alliance between robotics and pedagogy, with its latent ambiguities and legitimate suspicions, is an emblematic theme.

A further goal and a new challenge for the human is at stake, whose development, through robotics, big data and the Internet of Things, brings with it the need for a critical and conscious view of machines, a policy and educational planning for the care of our common home.

2   Technocratic Paradigm and New Humanism: Awareness, Education and Responsibility

We decide to give meaning and sentimental value to machines (if we get attached to our old car is more likely that we can love a human-like robot by projecting ancestral parental instincts onto it). In the same way we are the ones who attribute to the machine the ability to see us as human beings and to establish a relationship with us. In reality, the machine has no relationship with us in the human sense of the term. (Cingolani & Metta, 2015, pp. 7–8, trans. by the authors)

It interacts, through complex artificial intelligence algorithms, talks, drives a car and makes small operational decisions, but it does not feel emotions, nor does it have any sentimental structure or personal status.

However, sophisticated technical products—such as smartphones or humanoid robots—are capable of mediating patterns of relationships and directing the interests of well-identified power groups. Economic choices that appear to be simply instrumental are actually intentionally related to the type of social life they are intended to promote. It is inconceivable nowadays to think of technology as something simply functional: "The technological paradigm has become so dominant that it would be difficult to do without its resources and even more difficult to utilize them without being dominated by their internal logic" (Francesco, 2015, n. 108). It tends to exercise hegemonic dominance over politics and economics, and the latter is willing to assume any technological progress in the key of profit maximization, without paying attention to any negative consequences for human beings. For many reasons, it is not a question of making refined distinctions between different economic theories, but rather of recognizing the actual centrality of a reductive and individualistic mainstream in the factual orientation of the economy. Researchers and managers who do not admit this clearly, support it with facts when they do not address in a systemic way the issues of the dignity of work, the growing inequality in the distribution of wealth and the continuing serious ecological crisis in relation to the rights of future generations. Through their behaviour they affirm that the main objective corresponds to the marginal increase in profits. "In fact, if the market is governed solely by the principle of the equivalence in value of exchanged goods, it cannot produce the social cohesion that it requires in order to function well" (Benedetto XVI, 2009, n. 35). "We fail to see the deepest roots of our present failures, which have to do with the direction, goals, meaning and social implications of technological and economic growth" (Francesco, 2015, n. 109).

It is not the power of technology that is the fundamental issue, it is the "loyalty of the person to other human beings: the loyalty, responsibility, and respect that establish the educational relationship between people" (Colicchi Lapresa, 1994, p. 110, trans. by the authors). It is necessary to be aware of a disorientation perhaps even greater than the one described by Marshall McLuhan in 1964, regarding the rapid rise of the electric medium: "The technique of electricity is in our midst and we are stunned, deaf, blind and dumb in the face of its collision with the technique of Gutenberg" (1964, p. 18). Through the digital revolution, technological pervasiveness is greatly increased. Byung-Chul Han notes how "we are reprogrammed, without fully understanding this radical paradigm shift. We are behind the digital *medium* which, acting below the level of conscious decision, decisively modifies our behavior, our perception, our sensitivity, our thinking, our living together. Today we become intoxicated with the digital *medium*, without being able to fully assess the consequences of such intoxication" (Han, 2017, p. 9). This blindness to the implications of the changes and the simultaneous daze represents an irreducible component of the crisis of civilization, which is taking the form of a great cultural, spiritual and educational emergency. It implies a "long path of renewal" (Francesco, 2015, n. 202), investing every level of economic, political and social life, as well as daily life itself. The need for a new humanism arises from a new organization in everyday life experience, which can be considered fundamentally narrative. The individuals, both narrating and narrated, and the interpersonal relationships they weave, are not exhausted on the plane of logical-formal knowledge and technical application; they are always crossed by a dynamism of action and interpret existence in the flesh. "The desert grows in amplitude because at the same pace as the physical–geological–geographical desert, the desert that everyone hides within themselves grows to a greater extent and at a higher speed, that is, the aridity of the soul, of the heart and even of the mind that leads to pursue its own short-term profit at the expense of others, of contemporaries" (Anelli, 2016, p. 10, trans. by the authors), and of those to come.

Among technocratic paradigms and humanoid robots (Cingolani & Metta, 2015), we need a development that is marked by "that extraordinary valorization of individuals, regardless of their age and other concrete determinations, accompanied, however, by an equally important valorization of what is outside the individual: the natural and social world" (Bertolini, 1994, pp. 31–32, trans. by the authors). We must care for the world around us, and have respect for the human, throughout our lives.

"To *respect*, literally, means to *look away*. It is a regard. In relating respectfully to others, one refrains from pointing one's gaze indiscreetly. Respect presupposes a detached gaze, a pathos of distance. Today, this gaze yields to a vision devoid of distance, which is typical of entertainment" (Han, 2017, p. 10). A society without respect, without the *pathos* of distance, results in sensationalism, disinterest in depth and indifference.

"The environment must be seen as God's gift to all people, and the use we make of it entails a shared responsibility for all humanity, especially the poor and future generations" (Benedetto XVI, 2010, n. 2). "An integral ecology is also made up of simple daily gestures which break with the logic of violence, exploitation and selfishness. In the end, a world of exacerbated consumption is at the same time a world which mistreats life in all its forms" (Francesco, 2015, n. 230), and greed. Roberto Cingolani and Giorgio Metta note in this regard:

There is a problem of education and social awareness, which not only serves to steer human behaviour in the right direction, but also to maintain a high level of attention to the dangers that are not intrinsic to a technology per se, but which can result from its misuse or unexpected effects. To give an example, no one would have ever imagined that a civil aircraft could be a weapon of mass destruction, but the tragic events of September 11, 2001 have shown that the irresponsible use of any technology makes it dangerous. … Education in the proper use of technologies presupposes an ethical culture, as well as a scientific-technological one, without which humanity is not able to manage the results of its knowledge. (Cingolani & Metta, 2015, p. 45, trans. by the authors)

In his address to participants at the plenary assembly of the Pontifical Academy for Life 2019, Pope Francis observed,

This is why global bioethics is an important front on which to engage. It expresses awareness of the profound impact of environmental and social factors on health and life. This approach is very in tune with the integral ecology described and promoted in the Encyclical Laudato si'. … The possibility of intervening on living material to orders of ever smaller size, to process ever greater volumes of information, to monitor – and manipulate – the cerebral processes of cognitive and deliberative activity, has enormous implications: it touches the very threshold of the biological specificity and spiritual difference of the human being. In this sense, I affirmed that the distinctiveness of human life is an absolute good. (Francesco, 2019)

The topic of emerging and converging technologies must be made the subject of programmatic and incisive attention by pedagogy, in dialogue with the hard sciences and the humanities. Artificial intelligence, big data, robotics and technological innovations in general must be employed in ways that contribute to the richness of human education, the service of populations and the protection of our common home, rather than the exact opposite. The inherent dignity of every human being must be placed tenaciously at the centre of reflection and action.

In this sense, it is useful to recognize that the phrase "pedagogy of artificial intelligence", although certainly effective to mark the challenges of a pedagogy aimed at reflecting radically on technologies, is ambiguous and may lead to misunderstandings. The term should not conceal the fact that the functional automatisms of a machine are far removed from the human prerogatives of knowledge and action, consciousness and intentionality.

We need an authentically sustainable development, not an avoidance of the commitment to a pedagogy of artificial intelligence, in the face of responsibility for future generations.

3   Towards a Pedagogy of Artificial Intelligence

Artificial intelligence and robotics are destined to be one of the fundamental issues that will cross—and are already revolutionizing—the twenty-first century. In the intrinsic complexity that concerns technological acceleration, some issues affect us, far from stereotypical "catastrophic" versions. What emerges more and more is that the most debated issues are related to the human use of technologies, to the value options underlying any instrumental action and ultimately, to the moral sphere of artificial intelligence. "As with any invention (from steam trains to gunpowder, from nuclear energy to biotechnology), it is human choices that we have to worry about, not machines. The challenge is above all ethical, not technological" (Malavasi, 2019, p. 113, trans. by the authors).

A pedagogy of artificial intelligence moves from the awareness of the personal responsibility of one's own behaviours and ideas, from the skilful management of the instruments in the possession of human beings, even in the complexity and scientific sophistication of machine learning and big data. Educating, even in the medical-health field, to take on the radical innovations of this era, means recognizing the issues that can profoundly affect the development of the personality, the acquisition of skills and competences, the dynamics of learning with and through artificial intelligence, on the general training of professionalism in the entire working life-cycle, on the ontological and ethical commitment for a good life, with and for each other in just institutions, between the development of robots and the care of people. Medicine, which has always been sensitive to this dialectic between the scientific-technological and humanistic dimensions of human knowledge, can constitute a fertile ground for dialogue with artificial intelligence pedagogy. At stake is the understanding of the meanings mediated by intelligent machines, in the reading and interpretation of representations of the digital world, in the definition of a shared, sustainable and inclusive axiological orientation, aimed at health and well-being.

The revolution of artificial intelligence and robotics concerns above all knowledge, the contents of knowledge and the search for meaning of the human adventure on this planet, even before referring to some mechanical functionality or technical tool; it is a full part of the way humanity has, today and in the near future, to look at itself and the epochal challenges that await it, while at the same time operating a profound reconceptualization of its own cognitive, epistemological and moral maps. The adventure of knowledge, between fears and desires, fears and hopes, meets the digitalization of the world and relates to it to change, evolve, develop, even in a formative sense. Human uniqueness, even in the face of technological advances that seem to overcome more and more specific personal abilities, remains the extraordinary faculty of signification of the world, of otherness and of oneself, and constitutes the ability to make sense of everything that exists, from the simplest inanimate objects to complex entities, abstract objects and philosophical ideas. This semantic capital, continuously growing both on a phylogenetic and ontogenetic level, based on the logical coherence of cognitions, cannot remain indifferent to the revolution of the infosphere and informatic systems, as well as to the modifications to the encyclopedia of knowledge and to the conceptual design itself of human knowledge. Note in this regard Luciano Floridi: "this assumes particular importance in the context of current information societies, in which the multiplication of information sources produces an exponential but often disorganized accumulation of data, not always capable of translating into a real semantic capital growth" (2019, p. 140). In an emblematic way, it is precisely by starting from the growing impact of artificial intelligence that we can deeply rethink the essence of global education (*Bildung*), which today takes place without interruption in the dense interactive network between knowledge, information and people, between algorithms and relationships.

A pedagogy of artificial intelligence, even in the medical field, must be aware of the mutual interconnection that involves each and every one in the development of self-education throughout the life span. Being human as informational organisms does not mean giving in to posthuman or utopian drifts according to a phantom cyborg version: on the contrary, it means recognizing the originality of the humanum that lives, works and learns precisely in natural and artificial, physical and virtual environments, online and offline, aware of the growing fusion of such contrasts in hybrid and complex contexts from an ontological and perceptive point of view, but no less willing to host ethical perspectives of care and educational responsibility. "What we have in mind is rather a more subdued, less sensational, yet more crucial and profound change in our under-standing of what a human being is. ... We began to conceive of ourselves as inforg, not through some biotechnological transformation of our body, but, more seriously and realistically, through the radical transformation of our environment and the agents that operate there" (Floridi, 2014, p. 112).

It is the very reality of the world that seems to welcome automa-tion and digitization as significant processes of the era we are living in, with a much greater influence on humanity and its meanings, the more it appears as an original dynamic compared to the transformations of the past, unique in its ontological core. The autonomy of artificial intelli-gence and robotics, real or presumed, developed according to different degrees of complexity, scientific questions and pedagogical reflection on the awareness of its continuous diffusion in multiple theoretical and appli-cation contexts. "Automation, therefore, not only as an engineering push to build machines and automata, but as a more comprehensive perspective of meaning and production of our reality in the making" (Accoto, 2019, p. 4, trans. by the authors). Between utopian salvific illusions and ideolog-ical foreclosures, the machine induces humanity to reflect on itself and on the evolving condition that unites them, understanding in more depth the potential, limits, opportunities and fragility of the technosphere, noting without prejudices contained, languages and symbols of digital. It would be extremely reductive to approach the semantic richness of artificial intel-ligence with a mere functional reading, although it remains within the logical set of instrumentalities available to human civilization; on the contrary, the mutual influence between becoming personal and digital programming does not end in easy dichotomies but rather seems to enrich the phenomenological panorama of today's anthropological reality, as never before.

"Every form of being and life solicits, in its critical phases, forms of expression that find their full realization only in a new medium" (Han, 2017, p. 55). Faced with the desire to express one's life in a free, creative and supportive way, the automatisms of the algorithms express a challenge and an educational possibility, favouring the training of the competent professional and the integral development of the person. A pedagogy of artificial intelligence studies the material culture of its own digital age, interprets the educational essence of robotics in its cognitive, emotional and value implications, in the perspective of educating the person also through the media and digital devices that intersect in the process of personal and community learning.

> Returning to the moral aspects of the question: the impression is that digital technological progress in the age of AI allows for a progressive invasion of the intimacy and emotionality of users. The ethical awareness of this problem and the attempt to progressively empower the user are only the beginning of a process of human protection. (Maffettone, 2021, p. 159, trans. by the authors).

The protection of the human referred to by Sebastiano Maffettone does not seem to allude only to the necessary legal and political guarantees, requests particularly felt by public opinion in Western democracies, but also opens up the possibility of a broadly formative discourse. Cultivating people's creative resources, their talents, safeguarding their moral dignity and pointing out the impassable limits is fully part of the tasks of democratic institutions, but accompanying and encouraging such a profound process of change is unthinkable without adequate reflection and planning. "This implies a universal ethical perspective attentive to the themes of creation and human existence, to that humanism of life, fraternal and supportive, which is aimed at guiding transformations—from nanotechnologies to robotics—at the service of the person, of his formation and its fully human fulfillment" (Malavasi, 2020, p. 134, trans. by the authors).

References

Accoto, C. (2017). *Il mondo dato. Cinque brevi lezioni di filosofia digitale*. Egea.

Accoto, C. (2019). *Il mondo ex machina. Cinque brevi lezioni di filosofia dell'automazione*. Egea.

Anelli, F. (2016). La natura come creazione e le responsabilità dell'uomo. In C. Giuliodori & P. Malavasi (Eds.), *Ecologia integrale. Laudato si'. Ricerca, formazione,* conversione (pp. 3–10). Vita e Pensiero.

Benedetto XVI. (2009). Encyclical letter *Caritas in veritate*.

Benedetto XVI. (2010). Message for the celebration of the World Day of peace. *If you want to cultivate peace, protect creation*.

Bertolini, P. (1994). La mia posizione nei confronti del personalismo pedagogico. In G. Flores d'Arcais (Ed.), *Pedagogie personalistiche e/o pedagogie della persona* (pp. 31–54). La Scuola.

Brooks, R. (2017). L'intelligenza artificiale dominerà le nostre vite? *Internazionale, 1232*, 52–61.

Cingolani, R., & Metta, G. (2015). *Umani e umanoidi. Vivere con i robot.* il Mulino.

Colicchi Lapresa, E. (1994). Persona e verità dell'educazione. In G. Flores d'Arcais (Ed.), *Pedagogie personalistiche e/o pedagogie della persona* (pp. 89–112). La Scuola.

Ferraris, M. (2005). *Dove sei? Ontologia del telefonino.* Bompiani.

Floridi, L. (2014). *The fourth revolution.* Oxford University Press.

Floridi, L. (2019). *The logic of information.* Oxford University Press.

Francesco (2015). Encyclical letter *Laudato si' On care for our common home.*

Francesco. (2019, February 25). *Address of his holiness Pope Francis to participants in the Plenary Assembly of the Pontifical Academy for Life.* Retrieved September 30, 2021, from https://www.vatican.va/content/francesco/it/speeches/2019/february/documents/papa-francesco_20190225_plenaria-accademia-vita.html

Greenfield, A. (2017). *Radical technologies.* Verso Books.

Guardini, R. (1965). *Das Erde das Neuzeit.* Grünewald.

Han, B.-C. (2017). *In the swarm: Digital prospects* (E. Butler, Trans.). MIT Press. (Original work published 2013)

Kuniavsky, M. (2010). *Smart things: Ubiquitous computing user experience design.* Elsevier.

Maffettone, S. (2021). I dati tra valore morale, sociale e politico. L'etica pubblica nell'era digitale. In M. Bertolaso & G. Lo Storto (Eds.). *Etica digitale. Verità, responsabilità e fiducia nell'era delle macchine intelligenti* (pp. 147–162). Luiss University Press.

Malavasi, P. (2019). *Educare robot. Pedagogia dell'intelligenza artificiale.* Vita e Pensiero.

Malavasi, P. (2020). *Insegnare l'umano.* Vita e Pensiero.

McLuhan, M. (1964). *Understanding media: The extension of man.* Penguin.

Ravasi, G. (2017). *Adamo, dove sei? Interrogativi antropologici contemporanei.* Vita e Pensiero.

Valentini, V., Dinapoli, N., & Damiani A. (2013). The future of predictive models in radiation oncology: From extensive data mining to reliable modeling of the results. *Future Oncology, 9*, 311–313.

Connectivity in the Virtual Office Space: Catalyst or Impediment to TMT Agility?

Ionela Neacsu, Marta M. Elvira,
Carlos Rodríguez-Lluesma⬭, and Elvira Scarlat

Abstract In today's fast-paced and uncertain business landscape, coping with technology developments, increased demand for innovative products and competitive shifts require top management teams (TMTs) to respond to these challenges with increasingly higher levels of agility. Given the recent rise of ICT in business communication, in this conceptual paper we build on the attention-based view of the firm to shed light on the impact

I. Neacsu
Rennes School of Business, Rennes, France
e-mail: ionela.neacsu@esc-rennes.com

M. M. Elvira · C. Rodríguez-Lluesma (✉)
IESE Business School—University of Navarra, Madrid, Spain
e-mail: clluesma@iese.edu

M. M. Elvira
e-mail: melvira@iese.edu

E. Scarlat
IE Business School—IE University, Madrid, Spain

M. Bertolaso et al. (eds.), *Digital Humanism*,
https://doi.org/10.1007/978-3-030-97054-3_12

of ICT in shaping TMT agility. We discuss how ICT can either enhance or impair TMT agility, and identify TMT- and firm-level contingencies that boost the interplay between the two concepts.

Keywords TMT agility · Information technology · Communication technologies · Decision-making

1 INTRODUCTION

TMTs play the pivotal role in leading organizations. Their tasks include identifying growth opportunities (Penrose, 1980), planning and organizing resources (Whitley, 1989), and making critical decisions regarding acquisitions (Nadolska & Barkema, 2014), research and development investments (Kor, 2006), organizational reorientations (Tushman & Rosenkopf, 1996) or new product launches (Boeker, 1997). The design and implementation of such strategic actions hinge on access to accurate, up-to-date information about the business environment (Eisenhardt, 1989). Nevertheless, accelerated social change (Rosa & Scheuerman, 2009), and, in particular, technology developments, increased demand for innovative products, novel regulation and competitive shifts (Schiavone, 2011; Tushman & Murmann, 2003) often make information "inaccurate, unavailable, or obsolete" (Bourgeois & Eisenhardt, 1988, p. 816). Coping with these high-velocity environments requires TMTs to identify weak change signals and respond to them (Brown & Bessant, 2003; Eisenhardt & Martin, 2000; Krotov et al., 2014; Sharifi & Zhang, 2001) with increasingly higher levels of agility.

However, TMT agility remains an underexplored area of research. This is surprising, given the TMT's centrality to firm strategy (Cho & Hambrick, 2006; Eggers & Kaplan, 2009) and how much research attention has been paid to related phenomena. Research streams relevant to understanding TMT agility exist, but they are disconnected and targeted at other levels of analysis. For instance, prior research has analyzed agility at the organizational level, as a feature that companies foster to succeed

e-mail: elvira.scarlat@ie.edu

in changing environments (Dove, 2002; Goldman et al., 1995; Lu & Ramamurthy, 2011), black-boxing the role of the TMT in this process. Other researchers have analyzed the way ICT affects team communication and performance (Bailey et al., 2012; Maznevski & Chudoba, 2000), yet we still have a limited understanding of how ICT shapes TMTs' strategic agendas relative to traditional communication (Ocasio et al., 2018) and of its effects on TMT agility. Mixed conclusions also predominate regarding the impact of communication technologies on decision-making in teams: some researchers find that ICT enhances responsiveness to uncertain conditions (Lu & Ramamurthy, 2011), and others conclude that it discourages team collaboration and leads to disagreement and conflict (Levina & Vaast, 2008).

The purpose of this conceptual paper is to shed light on how ICT shapes TMT agility. To do so, we build on the attention-based view (ABV) theory (Ocasio, 1997, 2011; Ocasio & John, 2005; Ocasio et al., 2018) and discuss how ICT directs TMTs' attentional orientation and agility, depending on specific characteristics of the TMT and of the organization's environment.

We contribute to existing literature in several ways. First, we extend ABV research by presenting a theoretical framework for how communication technologies influence the attentional engagement of TMTs, thereby facilitating or impeding TMT agility. Second, we contribute to TMT research by developing and analyzing the notion of TMT agility, which we propose as a determinant of organizational agility. Third, we propose and analyze a series of factors that frame the relation between ICT and TMT agility. Specifically, based on Eisenhardt (1989), we identify and discuss the following factors: TMT interdependence; homogeneity; polychronicity; technological inclination; managerial competency; and firm internationalization. Finally, we show that the particular conditions of the TMT and the firm's environment are critical in using ICT for successful decision-making. In the last section of this paper, we describe the implications of our analysis for theory and practice, and propose paths that future research may take in order to advance this research.

2   TMT AGILITY AND ICT CAPABILITIES

An Attention-Based View of TMT Agility

TMTs constitute the "dominant coalition" (Cyert & March, 1963) of senior executives, who provide the interface between the firm and its environment (Hambrick et al., 2005) by making strategic decisions (Amason, 1996; Amason & Sapienza, 1997; Carpenter & Fredrickson, 2001; Hambrick et al., 1996; Simons et al., 1999). A TMT's distinctive knowledge base and degree of flexibility and responsiveness to changing environments will inherently affect the team's decision-making process. This implies that agility is critical for TMTs.

Since strategic decisions are usually higher in importance, urgency, complexity, relevance and uncertainty than those made by lower-level organizational actors (Amason, 1996; Clark & Maggitti, 2012; Eisenhardt, 1989), TMTs' tasks and skills differ from those of other organizational teams. TMTs' tasks are highly fluid and changeable, mostly unstandardized, context specific and interdependent (Whitley, 1989). Therefore, we need to distinguish TMT agility from its organizational counterpart—which has received most of the attention in previous research.

Extensive research has been conducted on how TMTs' decision-making ability influences firms' success (Hambrick & Mason, 1984; Raes et al., 2011). In this sense, TMT agility is a precondition for organizational agility: if TMTs are the catalyst for change in the organization (Hambrick & Mason, 1984), facilitating their agility is essential in a fast-paced business environment—and a first step in this process is understanding which factors enable it.

A TMT's attention to various stimuli shapes the way in which its members collaborate, share information, and act, ultimately influencing TMT agility. Simon (1947) introduced the concept of attention as a limited capability whose allocation shapes decision-making. Continuing his reasoning, Ocasio (1997) proposes that decision makers' actions depend on what issues and answers they focus their attention on. TMTs' attentional orientation depends on the objective reality that managers face at any given time (Cialdini et al., 1990), and managerial attention to environmental stimuli influences the strategic response of the organization (Strandholm et al., 2004). Thus, attention and communication structures connect individual information processing with the organizational context.

Ocasio (1997) advances three premises that explain attention distribution. First, decision makers' actions depend on their focus of attention. Individual attentional processes focus the cognitive efforts on a limited set of issues at any given time, to the detriment of other issues. Focused attention can be either controlled by the individual (i.e., an individual's intentional and mindful effort), or automatic, in which case environmental stimuli directly attract the individual's attention and trigger their action. Second, the focus of attention depends on the rules and norms particular to the situation in which decision makers find themselves. Third, this specific situation depends on how decision makers' attention is distributed between social relationships, communication, activities and procedures.

In line with the ABV theory, we define attention as "the noticing, encoding, interpreting, and focusing of time and effort by organization decision-makers" (Ocasio, 1997, p.189). Although Ocasio's (1997) focus is on organizational attention, he acknowledges that decision makers' attention influences their actions. Other researchers have also studied how ABV applies specifically to top management teams and concluded that TMTs' attention affects firm performance (Cho & Hambrick, 2006; Li et al., 2013).

ABV research generally studies how environmental stimuli that trigger strategic change affect attention engagement (Barnett, 2008; Joseph & Ocasio, 2012). In a recent review of the literature, Ocasio et al. (2018) note that while the traditional ABV underlines the importance of the information-processing capacity and the structural distribution of attention, future research should focus on how communication practices shape TMT attention. The authors propose that communication channels represent not just a measure of attention allocation, but arenas for sharing opinions, developing solutions, and brainstorming ideas, affecting managers' attention and actions (Cornelissen et al., 2015). As such, Ocasio et al. (2018) recommend examining how ICT generates, directs and transforms decision makers' attentional engagement, particularly in geographically distributed organizations. Our conceptual model responds to this call for research.

We propose that when technological developments define top managers' work and interactions, it is likely that managerial attention will be shaped by ICT (Cho & Hambrick, 2006; Eggers & Kaplan, 2009).

Hence, it is only natural that we raise the following question: how do ICT affect TMTs' decision-making and, consequently, their agility? We turn to this issue in the following section.

The Impact of ICT on TMT Agility

Digital systems have produced time- and location-flexible work environments, challenging the classical concept of the office as physical space (Chung et al., 2014). According to Gillam and Oppenheim (2006, p. 160), "networks, relationships and globalization typify this era. ... Electronic space, which coexists with geographical space, must be managed in order to maximize the opportunities it offers." The creation of this geographic dispersion–digital space coupling has led many teams to become partially or completely virtual (Cohen & Gibson, 2003; Kirkman et al., 2012). Most teams now communicate and coordinate through e-mail, messaging and web-conferencing applications without needing to be in the same place (Bailey et al., 2012; Maznevski & Chudoba, 2000).

In high-velocity environments, where constant change requires rapid response and flexibility, efficient communication (Smith et al., 1994) is critical to the strategic, high-stakes decisions TMTs make. Recent ABV research has evolved to propose communication not simply as a measure of relative attention (Cho & Hambrick, 2006; Tuggle et al., 2010), but as a factor that itself affects attention distribution (Ocasio, 2011). The specific content and practices of communication allow decision makers to jointly decide upon strategic changes in initiatives and actions (Ocasio et al., 2018). Designing and applying these strategies require attentional engagement, or a mindful and intentional allocation of cognitive resources towards planning, problem-solving and decision-making (Ocasio, 2011; Ocasio et al., 2018). Subject to the resulting allocation of cognitive resources, communication practices—and ICT in particular—may either enable or constrain TMT agility. Next, we discuss each of these alternatives.

ICT as Catalyst for TMT Agility

In many ways, ICT allows teams to replace in-person meetings with virtual interactions. While face-to-face meetings used to be seen as the standard practice in decision-making (Pinto et al., 1993), they have shortcomings. Research shows that in physical meetings, the risks associated with taking a minority position relative to other team members are higher than in virtual settings (Tan et al., 1998). Conversely, ICT is likely to shift participants' attention towards the most meritorious ideas, irrespective of who their proponents are. By directing the team's controlled attention and intentional cognitive efforts towards the best solutions, ICT may facilitate complex problem-solving and high-stakes decision-making (Ocasio, 2011), thereby providing a more levelled, agile communication platform.

ICT could free up meeting time by allowing TMT members to contact each other briefly whenever necessary via enterprise social networks, video-chat applications or e-mail, instead of allocating long time slots to discuss all issues at once. Research has also found that new information—critical for TMT agility—is often ignored in face-to-face meetings (Stasser & Titus, 1987). ICT could help overcome this impediment by facilitating quick, timely communications to TMT members, mitigating the challenges imposed by task complexity (Kock & Lynn, 2012) and facilitating the connections for adaptive enterprises (Haeckel, 1999) via digital systems (Sambamurthy et al., 2003).

Two aspects of team communication facilitate fast decision-making: informality (i.e., spontaneous conversations and unstructured meetings) and frequency (i.e., the amount of interaction among team members) (Smith et al., 1994). ICT can help timely brainstorming and joint development of new ideas by TMTs. For instance, chatting applications (e.g., Slack) allow team members to share updates with each other casually at will. Videoconferencing tools allow distributed teams to meet frequently and informally so that team members may stay aligned and informed. In sum, ICT may focus TMT members' attention towards urgent and important issues by enabling real-time communication (Chae ct al., 2014; Gillam & Oppenheim, 2006; Gilson et al., 2015; Hertel et al., 2005; Lu & Ramamurthy, 2011; Lucas & Olson, 1994).

Virtual information distribution through ICT can also enhance agility via reduced social isolation (Gillam & Oppenheim, 2006) and better task distribution between TMT members. For instance, e-mail communications could help scheduling and storage of individual and team information. Likewise, videoconferences may facilitate fast exchanges of information, building trust and close relationships between TMT members, and live chatting applications encourage cheap, spontaneous and informal conversations. Other groupware technologies, such as forums or shared networks, encourage collaboration and transparent information sharing, breaking down communication barriers (Gillam & Oppenheim, 2006).

Research has found that ICT reduces social loafing and increases perception of other team members' competence and satisfaction (Gilson et al., 2015). As a result, in recent decades, researchers have tried to conceptualize and test explanations of the formation and evolution of agile teams and organizations in the context of new ICT. This research concludes that if ICT helps TMTs design a management system that adopts and implements "routines" of agility (Winby & Worley, 2014), it can facilitate a more timely and effective response to changing circumstances.

ICT as Impediment to TMT Agility

While, as shown in the previous section, ICT may foster TMT agility, it may also block it (Overby et al., 2006; Weill et al., 2002), becoming a double-edged sword (Lu & Ramamurthy, 2011).

First, disagreeing on ICT usage can lead to suboptimal compromises, poor strategic decisions and likely poor TMT agility (Bailey et al., 2012). ICTs offer a wide range of communication tools differing in richness, channels and collaboration synchronicity (Daft & Lengel, 1984; Lu & Ramamurthy, 2011; Riopelle et al., 2003). As a result of their often different views and experiences, TMT members may differ in their preference for, and ability to use, various means of communication. To illustrate, some TMT members may prefer to discuss issues solely in face-to-face meetings, which may impede TMTs' access to real-time information, constraining agility.

Similarly, different opinions regarding the functionality of ICT media may lead TMT members to disagree about the utility of its features (Leonardi, 2011). Moreover, the choice of certain ICT tools may lead to different outcomes. For instance, stimuli from ICT notifications may involuntarily direct the attention of TMT members towards timely, albeit sometimes irrelevant information, shifting their attentional engagement and distracting them from dealing with important issues. Also, depending on the situation, using e-mail communication instead of videoconferencing tools could hamper information transmission due to the lack of non-verbal communication cues (Leonardi, 2010). In the absence of non-verbal cues, the likelihood of conflict increases, impeding communication

(Cramton, 2001) and leading to negative consequences in high-risk circumstances.

The relatively fixed characteristics of ICT (Allen & Boynton, 1991; Galliers, 2007; Lucas & Olson, 1994; Overby et al., 2006; Weill et al., 2002) can also generate disruption in information exchange, more frequent misunderstandings and lack of message coherence. For example, communication tools used for virtual meetings may shift participants' focus of attention by notifying them of an issue irrelevant to the meeting, but relevant to some other area of their responsibility. In addition, information overload caused by the availability of ICT media may lead TMT members to feel confused and absent, contributing to poor strategic decision-making. Kahneman (2011) suggests that information flooding may also lead to interminable debates and an exaggerated appreciation of empirical evidence over intuitive thinking.

A further potential issue may arise from the taken-for-grantedness of ICT. The fact that information *can* be distributed at any time does not mean that it *will* be, but TMT members may expect all the relevant information to be available. However, TMT members may fail to provide access to critical data on which all members depend (Levina & Vaast, 2008) and even fail to respond to TMT members' messages (Cramton, 2001). When team members are forced to interact via virtual media, they are also more likely to encounter trust issues (Bailey et al., 2012). Lack of trust among team members poses difficulties for their interaction, discouraging information sharing and collaboration (Jarvenpaa & Leidner, 1999), all of which impede TMT agility.

While the above studies provide ample evidence of the relation between ICT and TMT, we propose that the mixed findings of previous research can be reconciled by the specific contexts in which TMTs operate. In the following section we discuss how specific TMT- and organization-level contingencies may help enhance the relationship between ICT and TMT agility.

3   The Link Between ICT Capabilities and TMT Agility: Exploring Contingencies

Based on ABV, we suggest that one reason for the mixed effects ICTs have on TMT agility may stem from contingencies that facilitate an efficient use of ICT for some top teams. Certain characteristics of communication (form, frequency, length, etc.) direct managers' attention to specific issues, leaving other matters unattended to (Ocasio et al., 2018). How TMTs engage with ICT leads to them paying attention to a specific set of issues, and this selection process determines why some teams thrive and others fail in highly digitalized environments.

Depending on the context, communication practices elicit attention processes, which in turn determine the allocation of tasks and activities, and trigger the way managers select the issues they pay attention to. Examples of such decisions could be: 'What information needs to be shared?' 'What ICT tool is the most efficient in this situation?' 'How long should the communication be?' To achieve agility, TMTs need to make such decisions promptly. Eisenhardt (1989) identified five levers to accelerate decision-making: two-tier advice process, multiple simultaneous alternatives, decision integration, real-time information and consensus with qualification.

We integrate ABV theory and Eisenhardt's (1989) findings to identify the set of relevant contingencies enhancing the relation between ICT and TMT agility (see Table 1). Both TMT- and organizational-level characteristics generate the specific situation confronting top managers when they make high-stakes decisions, as we discuss below.

First, centralized power drains information from the common pool, hindering decision-making. The opposite happens in a two-tier advice process, where more than one person attends to an issue. Similarly, having multiple simultaneous alternatives enhances fast decision-making by allowing managers to shift rapidly between options if the first solution does not work (Eisenhardt, 1989). When TMT members work

Table 1 Dimensions of fast decision-making (Eisenhardt, 1989) and selected contingency factors

Fast decision-making dimension	Related contingency factor
Real-time information	TMT polychronicity TMT technological inclination
Multiple simultaneous alternatives	TMT interdependence Firm internationalization
Two-tier advice process	TMT heterogeneity
Consensus with qualification	TMT managerial competency
Decision integration	TMT interdependence TMT heterogeneity

closely together and their tasks are interrelated, they exchange information efficiently and engage in critical interactions. This pattern, called *TMT interdependence* (Hambrick et al., 2015), is likely to enhance brainstorming and allow TMT members to consider multiple alternatives creatively. We therefore expect TMT interdependence to relate to both the two-tier decision-making process and to the generation of multiple simultaneous alternatives.

Second, TMTs at international firms possess deeper experience and network ties (Ancona & Caldwell, 1992), allowing them to obtain information from a variety of sources and develop various solutions to problems. As a result, we expect *firm internationalization* to also enhance fast decision-making by developing multiple simultaneous alternatives.

Third, decision integration consists of merging multiple strategic decisions with one another and developing alternative plans (Eisenhardt, 1989). Considering various opinions and agreeing on a solution for each potential issue requires a homogeneous TMT group, as we explained in more detail in Sect. 2. We therefore argue that TMT *homogeneity* is an important contingency factor for our model.

Fourth, concerning the time dimension, Eisenhardt (1989) finds that firms that rely on real-time information incurred little or no time lag between event occurrence and reporting, and made faster decisions. Hambrick and Mason (1984) find that polychronous teams (i.e., those whose members engage in multiple tasks at once) share timely and relevant information informally, most of the time spontaneously. We argue then that real-time information sharing is enhanced by TMT *polychronicity*. Also, the success of TMTs with high technological inclination—understood as TMT members' ease in adopting new communication technologies—likely depends on their ability to produce and share real-time information. Thus, we propose that TMTs' *technological inclination* facilitates real-time information sharing.

Finally, consensus with qualification refers to taking an active approach to decision-making by involving all team members, with diverse experiences and knowledge, in the process. The quality of these decisions ultimately depends on managerial competency, understood as TMT's key abilities and potential capabilities (Furnham, 1990). To make fast decisions, TMT members use their superior skills to reach a consensus (Eisenhardt, 1989). Therefore, we propose that *managerial competency* facilitates TMTs reaching consensus and agile decision-making.

**The Link Between ICT Use and TMT Agility:
Exploring Contingencies**

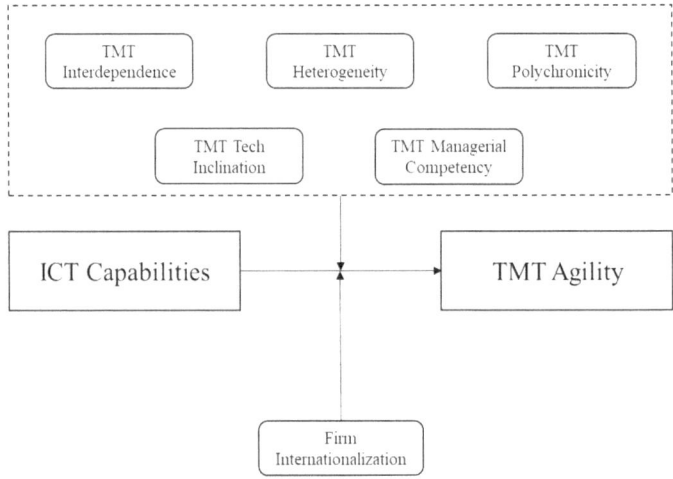

Fig. 1   The link between ICT use and TMT agility: exploring contingencies

The following sections consider in depth the proposed interaction effects (see Fig. 1).

TMT Interdependence

TMT members need to coordinate their work to meet their common targets. For teams to achieve more of their goals and improve firm performance (Barrick et al., 2007; Gully et al., 1995), effective communication is critical. This is especially true in high-velocity environments, where TMT members need to make high-stakes decisions quickly (Smith et al., 1994).

Barrick et al. (2007) define team interdependence as the extent to which members rely on one another to complete projects and tasks. The authors propose that efficient coordination efforts may result from an improvement in TMT members' behaviours and attitudes when performing tasks that require them to cooperate, learn from each other and motivate one another. They conclude that TMTs with high interdependence had higher team and firm performance when the team had more communication. Since ICT facilitates frequent communication, we expect it to improve the operational aspects of TMT decision-making, transactions, negotiations and information sharing in interdependent teams. By allowing TMT members to rapidly access and exchange information, schedule last-minute meetings or organize and synchronize agendas, we propose that in interdependent teams, ICT improves TMT agility.

Proposition 1 The impact of ICT use on TMT agility will be positively moderated by the degree of interdependence among TMT members.

TMT Homogeneity

We define TMT homogeneity as the extent to which TMT members are demographically and cognitively similar (Simons et al., 1999). Researchers have studied the impact of cultural differences on a firm's strategic decisions (Barkema & Shvyrkov, 2007; Carpenter, 2002; Hambrick & Mason, 1984; Hambrick et al., 1996; Heyden et al., 2015; Miller et al., 1998). The literature shows that team demographic homogeneity can reduce emotional conflicts, enhancing team agility. Similarly, shared knowledge and processes among team members improve decision-making and employee satisfaction (Raes et al., 2013; Standifer et al., 2015). TMT member homogeneity implies similar learning curves and experiences, which likely results in ICT platforms being similarly understood and adopted by all TMT members. TMT members' preference for similar communication methods may improve communication and facilitate achieving consensus. Thus, we expect that homogeneity among TMT members improves TMT communication when working together and enhances the relation between ICT and TMT agility.

Proposition 2 The impact of ICT use on TMT agility will be positively moderated by the degree of demographic homogeneity of the TMT.

TMT Polychronicity

Polychronicity is "the extent to which team members mutually prefer and tend to engage in multiple tasks simultaneously or intermittently instead of one at a time" (Souitaris & Maestro, 2010, p. 652). From an attention-based perspective, TMT polychronicity can be interpreted as an attention structure that leads to unscheduled interpersonal interactions rather than planned tasks (Souitaris & Maestro, 2010).

Decisions made by polychronous teams revolve around constant, continuous communication of timely information. Hambrick and Mason (1984) find that members of such teams share timely and relevant information informally, most of the time spontaneously. These teams obtain an information advantage via timely and relevant informal interactions between team members, most of the time casual rather than planned, which expedite decision-making (Eisenhardt, 1989) and promote agility. Since ICT facilitates such rapid and spontaneous exchanges of information, we propose that TMT polychronicity has a positive impact on the relationship between ICT and TMT agility.

Proposition 3 The impact of ICT use on TMT agility will be positively moderated by the degree of TMT polychronicity.

TMT Technological Inclination

We understand technological inclination as the ease with which TMT members become familiar with new technologies. Research reveals that while teams in general have increasingly adopted and used ICT (Daniels et al., 2001), TMTs in particular have been slower to do so (Lederer & Mendelow, 1988), partly due to their members' lack of familiarity with technology. Lederer & Mendelow's, 1988 survey found that TMT members were reluctant to gain the necessary knowledge for an efficient use of ICT tools. Things have not changed much today: when asked how prepared they are to meet the demands of a digitally disrupted business environment, more than 80% of executives responded that they felt underprepared (IMD, 2017).

This lack of familiarity with new technologies could be explained by TMT members' daily tasks. As discussed earlier, given that they constantly have to respond to urgent matters and make high-stakes decisions, top managers might have limited time to keep up with the high-paced evolution of new technologies. Naturally, managers' attention to ICT is higher in firms in technological industries (Eggers & Kaplan, 2009). Also, Marcel (2009) shows that managers' age affects their flexibility in information search and interpretation, and determines the manager's openness to innovation and reaction to change.

Research shows that the relative differences among team members in speed of accessing information via computer-mediated communication determines the speed of the firm's operations (Cramton, 2001). For instance, the author explains that when some team members rarely check their e-mail, the lack of interaction between team members slows the pace of the entire team, thus impairing its agility. Therefore, we expect TMT technological inclination to enhance the relation between ICT and TMT agility.

Proposition 4 The impact of ICT use on TMT agility will be positively moderated by the degree of TMT technological inclination.

TMT Managerial Competency

The impact of managerial characteristics and experience on strategic decision-making has been extensively analyzed in the context of the TMT literature, showing that firm behaviour is ultimately shaped by the quality of managerial decisions (Latukha & Panibratov, 2015; Ridge et al., 2015). Researchers have defined competency as the behavioural and personal characteristics, key abilities and potential capabilities that influence efficiency (Furnham, 1990; Menz, 2012).

TMT competency has a high cognitive dimension and comprises "a group's total pool of task-related skills, information, and perspectives" (Simons et al., 1999, p.663). In the context of TMT behaviour, competency consists of dimensions such as the degree to which team decisions improve organizational performance, the members' ability to make team decisions and collaborate in the future, and the extent to which team processes satisfy members' needs for growth and satisfaction (Latukha & Panibratov, 2015; McClelland & Brodtkorb, 2014).

Bantel and Jackson (1989) found that executives' prior experiences make them more flexible, and help them develop creative cognitive approaches to problem-solving. Clark and Maggitti (2012) find that experienced teams are better equipped to deal with uncertainty and ambiguity in highly competitive environments. TMT competency leads to a higher degree of cohesion between team members, translating into their ability to commit to mutual goals, efficient role distribution and leadership skills (Carpenter et al., 2001), and producing superior results when dealing with complex problems. These findings connect the concept of TMT competency to TMT agility.

At the same time, the need for competent TMTs to stay up to date likely incentivizes them to quickly adopt new communication technologies for instant information sharing. Thus, members of highly competent TMTs are more likely to see the potential for agility provided by ICT. We propose that managerial competency will positively influence the relation between ICT capabilities and TMT agility:

Proposition 5 The impact of ICT use on TMT agility will be positively moderated by the degree of managerial competency.

Firm Internationalization

A prerequisite of TMT agility is access to valuable information. Research has shown that TMTs of international firms have a broad understanding of international environments. They have large external networks and relational capital (Hitt et al., 2002), which offer a critical source of information and shapes TMT members' views about major organizational issues. Importantly, the vast set of experiences, knowledge and perspectives of international TMTs (Ancona & Caldwell, 1992) help top managers spawn creative, novel solutions to complex issues in a timely manner (Sharfman & Dean, 1997). TMTs with international exposure develop the flexibility needed to reconcile the different perspectives that are likely to occur when making high-stakes decisions (Arino & Reuer, 2002; Lee & Park, 2008).

Jointly taken, these features allow TMTs with international exposure to adjust quickly to changes in the environment, particularly in settings of high uncertainty (Lee & Park, 2008). In this sense, they possess the skills and cognitive characteristics to realize the potential that new communication technologies represent for their agility. At the same time, for geographically dispersed organizations, ICT provides an essential communication platform that allows instant communication. Therefore, we propose that TMTs with members based in different locations are more likely to benefit from ICT.

Proposition 6 TMT agility is more likely to benefit from the use of ICT when the firm operates in international markets.

4   Discussion and Conclusions

To date, researchers have analyzed TMT strategic decision-making by exploring the impact of team homogeneity and environmental conditions on TMT and firm output. Despite the barriers and challenges highlighted by previous work on the impact of digital capabilities on TMT decision-making, research remains divergent on the dual role of ICT capabilities, as they can both facilitate and obstruct TMT agility (Bergiel et al., 2008; Rosen et al., 2007). In this paper, we address this issue by exploring two main questions: (1) does the use of ICT capabilities enhance or reduce TMT agility? and (2) under what conditions is the impact of ICT capabilities on TMT agility more likely to be positive?

Regarding our first question, we build on previous theoretical and empirical work derived from the ABV theory to study how the ICT used at the TMT level shapes attention and agility. Based on earlier research, we explain that ICT can be a double-edged sword: on the one hand, ICT capabilities can enhance TMT agility, as they facilitate communication, increase work volume and division, reduce social isolation, and improve responsiveness to uncertain conditions and decision-making speed. On the other hand, ICT can impair TMT agility by disrupting information exchange and creating more frequent misunderstandings among team members.

Our second research question helps distinguish between these opposites according to specific contingency factors likely to affect the relation between ICT and TMT agility. To select these factors, we integrate the ABV approach with Eisenhardt's (1989) fast decision-making conditions. We discuss how TMT interdependence, TMT homogeneity, TMT managerial competency and TMT polychronicity, as well as firm-level contingencies such as the degree of firm internationalization, enhance the relation between ICT and TMT agility.

Implications for Theory and Practice

Our propositions add novel insight to the literature on organization and TMT decision-making. We present a theoretical framework to understand the relation between the use of ICT and its impact on TMT agility, and integrate research on the conditions that may enhance this relation. This approach advances our understanding of the role played by ICT systems in providing organizations with a platform that enables collaboration between TMT members, regardless of their geographic location and availability.

Our theoretical framework has several implications for research on TMTs and ICT. First, we conceptualize the notion of agility in the context of TMTs by discussing its multiple dimensions and its importance for strategic decision-making. Previous research has studied agility as a general organizational trait, regarding it a source of competitive advantage (Brown & Bessant, 2003; Dove, 2002; Goldman et al., 1995; Horney et al., 2010; Krotov et al., 2014; Lu & Ramamurthy, 2011; Sharifi & Zhang, 2001; Winby & Worley, 2014). In this study, we emphasize the way TMTs focus their attention on set goals, develop corresponding strategies and commit to their execution. We explore TMTs' ability to respond to changes and uncertainty by keeping a balance between the implementation of already existing managerial models and the adoption of more novel and efficient managerial practices.

Second, we build on previous theoretical and empirical work on digital capabilities to highlight the role of ICT as a critical input for firm activity (Gillam & Oppenheim, 2006; Gilson et al., 2015; Hertel et al., 2005). In particular, we analyze the role ICT plays in altering TMT virtuality and cohesion by exploring how ICT capabilities enable TMT agility (such as the capacity to speed up decision-making, team communication and its responsiveness to uncertainty), as well as its role in simplifying task complexity and reducing isolation through the removal of spatial barriers. By focusing specifically on TMT agility, our arguments add nuance to previous research on the more general impact of technology on team output.

Many companies recognize the benefits of using ICT appropriately and thus are increasingly moving towards virtuality. Take for example GitLab, a software development start-up: the company has no headquarters and everyone, including the CEO, works remotely. The company's co-founders, Sid Sijbrandij and Dmitriy Zaporozhets, lived in the Netherlands and Ukraine respectively, and use Skype to communicate. This trend towards virtuality illustrates the urgent need to understand the impact of ICT on TMTs' decision-making processes.

Given the nature of a top executive's job, which prioritizes decision-making efficiency, ICT can facilitate communication among TMT members. Interestingly, top executives are among the slowest groups to adopt virtuality: their schedules are notoriously filled with face-to-face meetings (Porter & Nohria, 2018), where most decisions are made. Despite the undeniable benefits of this kind of meeting, they might not need to be so frequent—anecdotal evidence is full of cases of executives stating that meetings consume a significant portion of their time. ICT can help TMT members communicate more efficiently, and our reasoning suggests that given the potential for ICT to help optimize TMTs' agendas, becoming proficient in virtuality is of paramount importance.

Limitations and Future Research

This conceptual study opens up discussion on an issue ever more prevalent among organizations, namely the role of ICT in shaping the attention of TMTs and influencing their agility. We propose that, as ICT becomes even more critical to strategic decision-making, its impact on TMT agility will be strongly influenced by contingencies from the internal and external environments.

Nevertheless, this paper leaves some points unanswered and suggests avenues for future research. First, as a conceptual piece, it does not empirically address the propositions. An empirical analysis could also identify the optimal frequency for ICT usage among team members and reveal additional contingency factors relevant for the studied relationship.

Also, we do not study separately the impact of the various types of technologies, and whether their effectiveness may lead to different outcomes in terms of TMT agility. Anecdotal evidence suggests that some tools are more efficient than others, but their direct impact is yet to be estimated. Research could next analyze whether TMTs need to keep a balance between following an established managerial model—based on physical meetings—and the novel usage of more recent ICT platforms, which facilitate instant information sharing for distributed team members in order to develop agility. We also encourage future research to further examine the impact of technology on the "millennial generation" (Gorman et al., 2004; Lu & Ramamurthy, 2011), arguing that the impact of ICT on millennials' behaviour may differ substantially relative to previous generations (Myers & Sadaghiani, 2010). Members of more recent generations may feel more comfortable using ICT (Morris & Venkatesh, 2000) and thus may be able to accomplish a wide range of tasks in a quicker and more efficient way, contributing to a more agile TMT. On the contrary, reluctance or fear of using ICT by members of previous generations may have a more negative impact on the agility of a TMT.

Lastly, as information accumulation becomes ubiquitous, many organizations and team members risk becoming distracted by the considerable amount of available information, and thus engage in biased decision-making. This draws attention to the importance of this research topic for better understanding managerial decision-making, and shows that further research on these potential side-effects is needed. Overall, empirical research should investigate what makes the impact of ICT on TMT agility more likely to be positive or negative.

References

Allen, B. R., & Boynton, A. C. (1991). Information architecture: In search of efficient flexibility. *MIS Quarterly, 15*(4), 435–445.

Amason, A. C. (1996). Distinguishing the effects of functional and dysfunctional conflict on strategic decision making: Resolving a paradox for top management teams. *Academy of Management Journal, 39*(1), 123–148.

Amason, A. C., & Sapienza, H. J. (1997). The effects of top management team size and interaction norms on cognitive and affective conflict. *Journal of Management, 23*(4), 495–516.

Ancona, D. G., & Caldwell, D. F. (1992). Demography and design: Predictors of new product team performance. *Organization Science, 3*(3), 321–341.

Arino, A., & Reuer, J. J. (2002). Contractual renegotiations in strategic alliances. *Journal of Management, 28*, 47–69.

Bailey, D. E., Leonardi, P. M., & Barley, S. R. (2012). The Lure of the Virtual. *Organization Science, 23*(5), 1485–1504.

Bantel, K. A., & Jackson, S. E. (1989). Top management and innovation in banking: Does the composition of the top team make a difference? *Strategic Management Journal, 10*(S1), 107–124.

Barkema, H. G., & Shvyrkov, O. (2007). Does top management team diversity promote or hamper foreign expansion? *Strategic Management Journal, 28*(7), 663–680.

Barnett, M. L. (2008). An attention-based view of real options reasoning. *Academy of Management Review, 33*(3), 606–628.

Barrick, M. R., Brandley, B. H., Kristof-Brown, A. L., & Colbert, A. E. (2007). The moderating role of top management team interdependence: Implications for real teams and working groups. *Academy of Management Journal, 50*(3), 544–557.

Bergiel, J. B., Bergiel, E. B., & Balsmeier, P. W. (2008). Nature of virtual teams: A summary of their advantages and disadvantages. *Management Research Review, 31*(2), 99–110.

Boeker, W. (1997). Strategic change: The influence of managerial characteristics and organizational growth. *Academy of Management Journal, 40*(1), 152–170.

Bourgeois, L. J., & Eisenhardt, K. M. (1988). Strategic decision processes in high velocity environments: Four cases in the microcomputer industry. *Management Science, 34*(7), 816–835.

Brown, S., & Bessant, J. (2003). The manufacturing strategy-capabilities links in mass customization and agile manufacturing—An exploratory study. *International Journal of Operations and Production Management, 23*(7), 707–730.

Carpenter, M. A. (2002). The implications of strategy and social context for the relationship between top team management heterogeneity and firm performance. *Strategic Management Journal, 23*(3), 275–284.

Carpenter, M. A., & Fredrickson, J. W. (2001). Top management teams, global strategic posture, and the moderating role of uncertainty. *Academy of Management Journal, 44*(3), 533–554.

Carpenter, M. A., Sanders, W. G., & Gregersen, H. B. (2001). Bundling human capital with organizational context: The impact of international assignment experience on multinational firm performance and CEO pay. *Academy of Management Journal, 44*(3), 493–511.

Chae, H., Koh, C., & Prybutok, V. (2014). Information technology capability and firm performance: Contradictory findings and their possible causes. *MIS Quarterly, 38*(1), 305–326.

Cho, T. S., & Hambrick, D. C. (2006). Attention as the mediator between top management team characteristics and strategic change: The case of airline deregulation. *Organization Science, 17*(4), 453–469.

Chung, S., Lee, K. Y., & Kim, K. (2014). Job performance through mobile enterprise systems: The role of organizational agility, location independence, and task characteristics. *Information and Management, 51*(6), 605–617.

Cialdini, R. B., Reno, R. R., & Kallgren, C. A. (1990). A focus theory of normative conduct: Recycling the concept of norms to reduce littering in public places. *Journal of Personality and Social Psychology, 58*(6), 10–15.

Clark, K. D., & Maggitti, P. G. (2012). TMT potency and strategic decision-making in high technology firms. *Journal of Management Studies, 49*(7), 1168–1193.

Cohen, S. G., & Gibson, C. B. (2003). *Virtual teams that work: Creating conditions for virtual teams effectiveness.* Jossey-Bass.

Cornelissen, J. P., Durand, R., Fiss, P. C., Lammers, J. C., & Vaara, E. (2015). Putting communication front and center in institutional theory and analysis. *Academy of Management Review, 40*(1), 10–27.

Cramton, C. D. (2001). The mutual knowledge problem and its consequences for dispersed collaboration. *Organization Science, 12*(3), 346–371.

Cyert, R. M., & March, J. G. (1963). A behavioral theory of the firm. *Englewood Cliffs, NJ, 2*(4), 169–187.

Daft, R. L., & Lengel, R. H. (1984). *Information richness: A new approach to manager information processing and organization design.* JAI Press.

Daniels, K., Lamond, D., & Standen, P. (2001). Teleworking: Frameworks for organizational research. *Journal of Management Studies, 38*(8), 1151–1185.

Dove, R. (2002). *Response ability: The language, structure, and culture of the agile enterprise.* Wiley.

Eggers, J. P., & Kaplan, S. (2009). Cognition and renewal: Comparing CEO and organizational effects on incumbent adaptation to technical change. *Organization Science, 20*(2), 461–477.

Eisenhardt, K. M. (1989). Making fast strategic decisions in high-velocity environments. *Academy of Management Journal, 32*(3), 543–576.

Eisenhardt, K. M., & Martin, J. A. (2000). Dynamic capabilities: What are they? *Strategic Management Journal, 21*(10–11), 1105–1121.

Furnham, A. (1990). Can people accurately estimate their own personality test scores?. *European Journal of Personality, 4*(4), 319–327.

Galliers, R. D. (2007). *Strategizing for agility: Confronting information systems inflexibility in dynamic environments.* Butterworth Heinemann.

Gillam, C., & Oppenheim, C. (2006). Review article: Reviewing the impact of virtual teams in the information age. *Journal of Information Science, 32*(2), 160–175.

Gilson, L. L., Maynard, M. T., Young, N. C. J., Vartiainen, M., & Hakonen, M. (2015). Virtual teams research 10 years, 10 themes, and 10 opportunities. *Journal of Management, 41*(5), 1313–1337.

Goldman, S., Nagel, R., & Preiss, K. (1995). *Agile competitors and virtual organizations.* Van Nostr and Reinhold.

Gorman, P., Nelson, T., & Glassman, A. (2004). The millennial generations: A strategic opportunity. *Organizational Analysis, 12*(3), 255–270.

Gully, S. M., Devine, D. J., & Whitney, D. J. (1995). A meta-analysis of cohesion and performance: Effects of level of analysis and task interdependence. *Small Group Research, 26*(4), 497–520.

Haeckel, S. H. (1999). *Adaptive enterprise: Creating and leading sense-and-respond organizations.* Harvard Business School Press.

Hambrick, D. C., Finkelstein, S., & Mooney, A. C. (2005). Executive job demands: New insights for explaining strategic decisions and leader behaviors. *Academy of Management Review, 30*(3), 472–491.

Hambrick, D. C., & Mason, P. A. (1984). Upper echelons: The organization as a reflection of its top managers. *Academy of Management Review, 9*(2), 193–206.

Hambrick, D. C., Cho, T. S., & Chen, M. (1996). The influence of top management team heterogeneity on firms' competitive moves. *Administrative Science Quarterly, 41*(4), 659–684.

Hambrick, D. C., Humphrey, S. E., & Gupta, A. (2015). Structural interdependence within top management teams: A key moderator of upper echelons predictions. *Strategic Management Journal, 36*(3), 449–461.

Hertel, G., Geister, S., & Konradt, U. (2005). Managing virtual teams: A review of current empirical research. *Human Resource Management Review, 15*(1), 69–95.

Heyden, M., Sidhu, J. S., & Volberda, H. W. (2015). The conjoint influence of top and middle management characteristics on management innovation. *Journal of Management, 44*(4), 1505–1529.

Hitt, M. A., Lee, H. U., & Yucel, E. (2002). The importance of social capital to the management of multinational enterprises: Relational networks among Asian and Western firms. *Asia Pacific Journal of Management, 19*(2–3), 353–372.

Horney, N., Pasmore, B., & O'Shea, T. (2010). Leadership agility: A business imperative for a VUCA world. *People and Strategy, 33*(4), 32–38.

IMD. (2017). *Redefining leadership for a digital age.* Retrieved 10 December 2020, from: https://www.imd.org/contentassets/25fdd7355de14eb3a157d 3b712222ef1/redefining-leadership

Jarvenpaa, S. L., & Leidner, D. E. (1999). Communication and trust in global virtual teams. *Organization Science, 10*(6), 791–815.

Joseph, J., & Ocasio, W. (2012). Architecture, attention, and adaptation in the multibusiness firm: General Electric from 1951 to 2001. *Strategic Management Journal, 33*(6), 633–660.

Kahneman, D. (2011). *Thinking, fast and slow.* Macmillan.

Kirkman, B., Gibson, C. B., & Kim, K. (2012). Across borders and technologies: Advancements in virtual teams research. In *The Oxford handbook of organizational psychology* (Vol. 2). Oxford University Press.

Kock, N., & Lynn, G. S. (2012). Lateral collinearity and misleading results in variance-based SEM: An illustration and recommendations. *Journal of the Association for Information Systems, 13*(7), 546–580.

Kor, Y. Y. (2006). Direct and interaction effects of top management team and board compositions on R&D investment strategy. *Strategic Management Journal, 27*(11), 1081–1099.

Krotov, V., Junglas, I., & Steel, D. (2014). The mobile agility framework: An exploratory study of mobile technology enhancing organizational agility. *Journal of Theoretical and Applied Electronic Commerce Research, 10*(3), 1–17.

Latukha, M. O., & Panibratov, A. (2015). Top management teams' competencies for international operations: Do they influence a firm's result? *Journal of General Management, 40*(4), 45–68.

Lederer, A. L., & Mendelow, A. L. (1988). Convincing top management of the strategic potential of information systems. *MIS Quarterly, 12*(4), 525–534.

Lee, H. U., & Park, J. H. (2008). The influence of top management team international exposure on international alliance formation. *Journal of Management Studies, 45*(5), 961–981.

Leonardi, P. M. (2010). Information, technology, and knowledge sharing in global organizations: Cultural differences in perceptions of where knowledge lies. In *Communication and organizational knowledge* (pp. 111–134). Routledge.

Leonardi, P. M. (2011). Innovation blindness: Culture, frames, and cross-boundary problem construction in the development of new technology concepts. *Organization Science, 22*(2), 347–369.

Levina, N., & Vaast, E. (2008). Innovating or doing as told? Status differences and overlapping boundaries in offshore collaboration. *MIS Quarterly, 32*(2), 307–332.

Li, Q., Maggitti, P. G., Smith, K. G., Tesluk, P. E., & Katila, R. (2013). Top management attention to innovation: The role of search selection and intensity in new product introductions. *Academy of Management Journal, 56*(3), 893–916.

Lu, Y., & Ramamurthy, K. (2011). Understanding the link between information technology capability and organizational agility: An empirical examination. *MIS Quarterly, 35*(4), 931–954.

Lucas, H. C., Jr., & Olson, M. (1994). The impact of information technology on organizational flexibility. *Journal of Organizational Computing and Electronic Commerce, 4*(2), 155–177.

Marcel, J. J. (2009). Why top management team characteristics matter when employing a chief operating officer: A strategic contingency perspective. *Strategic Management Journal, 30*(6), 647–658.

Maznevski, M. L., & Chudoba, K. M. (2000). Bridging space over time: Global virtual team dynamics and effectiveness. *Organization Science, 11*(5), 473–492.

McClelland, P. L., & Brodtkorb, T. (2014). Who gets the lion's share? Top management team pay disparities and CEO power. *Journal of General Management, 39*(4), 55–73.

Menz, K. (2012). Corporate governance and credit spreads—Correlation, causality, or neither nor? *International Review of Applied Financial Issues and Economics, 4*(1), 1–10.

Miller, C. C., Burke, L. M., & Glick, W. H. (1998). Cognitive diversity among upper-echelon executives: Implications for strategic decision processes. *Strategic Management Journal, 19*(1), 39–59.

Morris, M. G., & Venkatesh, V. (2000). Age differences in technology adoption decisions: Implications for a changing work force. *Personnel Psychology, 53*(2), 375–403.

Myers, K., & Sadaghiani, K. (2010). Millennials in the workplace: A communication perspective on millennial's organizational relationships and performance. *Journal of Business and Psychology, 25*(2), 225–238.

Nadolska, A., & Barkema, H. G. (2014). Good learners: How top management teams affect the success and frequency of acquisitions. *Strategic Management Journal, 35*(10), 1483–1507.

Ocasio, W. (1997). Towards an attention-based view of the firm. *Strategic Management Journal, 18*(S1), 187–206.

Ocasio, W. (2011). Attention to attention. *Organization Science, 22*(5), 1286–1296.

Ocasio, W., & John, J. (2005). An attention-based theory of strategy formulation: Linking decision making and guided evolution in strategy processes. *Advances in Strategic Management, 22*, 39–61.

Ocasio, W., Laamanen, T., & Vaara, E. (2018). Communication and attention dynamics: An attention-based view of strategic change. *Strategic Management Journal, 39*(1), 155–167.

Overby, E., Bharadwaj, A., & Sambamurty, V. (2006). Enterprise agility and the enabling role of information technology. *European Journal of Information Systems, 15*(2), 120–131.

Penrose, E. T. (1980). *The theory of the growth of the firm* (2nd ed.). M. E. Sharpe.

Pinto, M. B., Pinto, J. K., & Prescott, J. E. (1993). Antecedents and consequences of project team cross-functional cooperation. *Management Science, 39*(10), 1281–1297.

Porter, M. E., & Nohria, N. (2018). How CEOs manage time. *Harvard Business Review, 96*(4), 42–51.

Raes, A., Heijltjes, M. G., Glunk, U., & Roe, R. A. (2011). The interface of the top management team and middle managers: A process model. *Academy of Management Review, 36*(1), 102–126.

Raes, A., Bruch, H., & De Jong, S. B. (2013). How top management team behavioral integration can impact employee work outcomes: Theory development and first empirical tests. *Human Relations, 66*(2), 167–192.

Ridge, J. W., Aime, F., & White, M. A. (2015). When much more of a difference makes a difference: Social comparison and tournaments in the CEO's top team. *Strategic Management Journal, 36*(4), 618–636.

Riopelle, K., Gluesing, J. C., Alcordo, T. C., Baba, M. L., Britt, D., McKether, W., Monplaisir, L., Ratner, H. H., & Wagner, K. H. (2003). Context, task, and the evolution of technology use in global virtual teams. In C. B. Gibson &

S. G. Cohen (Eds.), *Virtual teams that work: Creating conditions for virtual team effectiveness* (pp. 239–264). Jossey-Bass.

Rosa, H., & Scheuerman, W. E. (Eds.). (2009). *High speed society: Social acceleration, power and modernity*. Pennsylvania State University Press.

Rosen, B., Furst, S., & Blackburn, R. (2007). Overcoming barriers to knowledge sharing in virtual teams. *Organizational Dynamics, 36*(3), 259–273.

Sambamurthy, V., Bharadwaj, A., & Grover, V. (2003). Shaping agility through digital options: Reconceptualizing the role of information technology in contemporary firms. *MIS Quarterly, 27*(2), 237–263.

Schiavone, F. (2011). Strategic reactions to technology competition: A decision making model. *Management Decision, 49*(5), 801–809.

Shah, A. (2019). The office of the future is no office at all, says startup. *The Wall Street Journal*. https://www.wsj.com/articles/the-office-of-the-future-is-no-office-at-all-says-startup-11557912601

Sharfman, M. P., & Dean, J. W. (1997). Flexibility in strategic decision making: Informational and ideological perspectives. *Journal of Management Studies, 34*(2), 191–217.

Sharifi, H., & Zhang, Z. (2001). Agile manufacturing in practice. *International Journal of Operations and Production Management, 21*(5/6), 772–795. https://doi.org/10.1108/01443570110390462

Simon, H. A. (1947). *Administrative behavior: A study of decision-making processes in administrative organization*. Macmillan.

Simons, T., Pelled, L. H., & Smith, K. A. (1999). Making use of difference: Diversity, debate, and decision comprehensiveness in top management teams. *Academy of Management Journal, 42*(6), 662–673.

Smith, K. G., Smith, K. A., Olian, J. D., Sims, H. P., Jr., O'Bannon, D. P., & Scully, J. A. (1994). Top management team demography and process: The role of social integration and communication. *Administrative Science Quarterly, 39*, 412–438.

Souitaris, V., & Maestro, B. M. M. (2010). Polychronicity in top management teams: The impact on strategic decision processes and performance of new technology ventures. *Strategic Management Journal, 31*(6), 652–678.

Standifer, R. L., Raes, A. M., Peus, C., Passos, A. M., Santos, C. M., & Weisweiler, S. (2015). Time in teams: Cognitions, conflict and team satisfaction. *Journal of Managerial Psychology, 30*(6), 692–708.

Stasser, G., & Titus, W. (1987). Effects of information load and percentage of shared information on the dissemination of unshared information during group discussion. *Journal of Personality and Social Psychology, 51*(July), 81–93.

Strandholm, K., Kumar, K., & Subramanian, R. (2004). Examining the interrelationships among perceived environmental change, strategic response, managerial characteristics, and organizational performance. *Journal of Business Research, 57*(1), 58–68.

Tan, B. C., Wei, K. K., Watson, R. T., & Walczuch, R. M. (1998). Reducing status effects with computer-mediated communication: Evidence from two distinct national cultures. *Journal of Management Information Systems, 15*(1), 119–141.

Tuggle, C. S., Sirmon, D. G., Reutzel, C. R., & Bierman, L. (2010). Commanding board of director attention: Investigating how organizational performance and CEO duality affect board members' attention to monitoring. *Strategic Management Journal, 31*(9), 946–968.

Tushman, M. L., & Murmann, J. P. (2003). Dominant designs, technology cycles, and organizational outcomes. In R. Garud, A. Kumaraswamy, & R. Langlois (Eds.), *Managing in the modular age: Architectures, networks, and organizations* (pp. 316–347). Blackwell.

Tushman, M. L., & Rosenkopf, L. (1996). Executive succession, strategic reorientation and performance growth: A longitudinal study in the US cement industry. *Management Science, 42*(7), 939–953.

Weill, P., Subramani, M., & Broadbent, M. (2002). Building IT infrastructure for strategic agility. *MIT Sloan Management Review, 44*(1), 57–65.

Whitley, R. (1989). On the nature of managerial tasks and skills: Their distinguishing characteristics and organization. *Journal of Management Studies, 26*(3), 209–224.

Winby, S., & Worley, C. G. (2014). Management processes for agility, speed, and innovation. *Organizational Dynamics, 43*(3), 225–234.

CHAPTER 13

Digital Is the New Dimension

Mei Lin Fung and Leng Leroy Lim

Abstract Digital technology connects the world but generates collisions that were avoided when distance separated people and ideas. Wealth meets poverty online, and Western science and economics meet Eastern body-mind-spirit. Understanding and empathy are torn by growing distrust. Yet digital technology also brings great benefits – enhanced food security, efficiencies in many fields and tools to steward our planet.

Digital accounting and tracking today primarily serve for-profit private enterprises, making a few people rich while constraining opportunities for far more. Divisions tilt the winnings to whichever tribe leverages digital technology to win the power game. This parallels how European nations used weapons technology and division to colonize and exploit indigenous peoples. To change this, we must avoid framing choices as binary, as a face-off between personal freedom and collective needs; instead, we need paths that respect diverse cultures and reconcile differences. Southeast Asia, where diverse peoples have learned to cooperate and support each

M. L. Fung (✉) · L. L. Lim
People Centered Internet, Palo Alto, CA, USA
e-mail: meilin@PeopleCentered.net

M. Bertolaso et al. (eds.), *Digital Humanism*,
https://doi.org/10.1007/978-3-030-97054-3_13

other, is one example that offers guidance on building trust in a digital world.

This is a time of great potential as well as danger. If digital technology can be managed for everyone's benefit, society will be much improved. We have an incredible opportunity to share and learn.

Keywords Digital · Technology · Potential · Danger · Global · Dimension

1   Introduction

Digital technology has enabled the world to be connected as never before. Those who were able to work digitally during the Covid-19 pandemic continued to enjoy social life, earning and learning, entertainment, food and family while minimizing the hazards of the global pandemic.

Digital connection crosses boundaries, enabling increasing transparency in work cultures and product and service reliability. Transparency also lets people peer across national borders; they can now "google" cities, fields and forests; listen through the walls of homes and offices; and visit universities, museums, libraries and archives at any time, no longer limited by opening hours and staffing. The mystery of new places and experiences used to require exploration in person. The digital dimension now exposes us (negative) and connects (positive) us to people we never knew existed. We are leaving our digital trails everywhere we go, not realizing who may be looking, unaware we might have new kinds of ownership rights.

Digital frontiers are merging with economics and science: behavioural, social, environmental and biological. What is public, what is private? Is there even a defined boundary any more? Deeply personal information can be known by others even before one is aware of it oneself. For example, AI bots can now identify gay or pregnant people with 80% accuracy even before that person might be aware of it themselves (Wang & Kosinski, 2018).

What are the new arenas in which we have to devise boundaries? We see, listen to and evaluate a huge mix of trivial, useful, bogus and real information that is overwhelming us, which we lack the capacity to process cognitively. How then can we create responsible channels, how can we come together to set boundaries appropriately while stewarding freedom of expression and independence for people in their desire to explore the world? Online porn viewing taking place in family homes is confronting parents of even the very young. What are permissible boundaries for trustworthy teenagers? What about those with no parental or adult oversight at all? Who is responsible for setting boundaries for child online safety? It's an open question in almost every society and country.

2   Boundaries

Historically, physical geography made for natural boundaries. Where natural boundaries like rivers and oceans did not exist, people created actual walls—like the Great Wall of China, or the fence in one's backyard. Boundaries can protect. Yet good boundaries have *gates* that allow trade and communication as cells in the body allow the right chemicals to flow in and out.

Technology today is evolving in ways that erase borders and boundaries, and yet the hoped-for and touted benefits are not being realized, or are realized unevenly or unfairly distributed. There is a growing backlash to techno-optimism. Like the colonial ships that sought spices for rich rewards, leading to conscious or unconscious ravaging of the indigenous people and lands, how much should we be concerned about the prospects of digital colonialism?

With bookstores, workplaces, and perhaps even playgrounds and restaurants changing, we barely take into account in our science, research and technological innovation that this is a bountiful and exciting time to be curious. We are living in an era in which people from vastly different backgrounds are able to come together to imagine how the future will unfold, and in the process discover new friends and build bonds.

Through social media, video and other digital images, previously separated people see each other's lives and often there are distorting lenses in the way. With super-rich indulgences now newsworthy, social injustice is literally *in your face* for the have-nots. Opportunities for scholarships and jobs bring together people from faraway places in ways that can both increase and decrease understanding. In fact, knowing more about the *other* can as easily provoke anxiety, hostility and paranoia. For example, China's rise from poverty is an achievement which is experienced by many Westerners as either bewildering or frightening or both, because "foreign otherness" is not assuaged by mere connectivity. What are needed, but missing, are new approaches to making friends and learning about each other in ways that do not make our differences the cause for conflict.

Only a few people, corporations and institutions manage the digital levers. They are the ones who possess the digital magic, enabled by maths, science, and economics, and honed by years of surviving at the top-of-the-food-chain corporations or the very tech savvy public servants in Estonia, Singapore, India and Switzerland. The vast majority of people in government, small and medium businesses and those working in civil society have been somewhat on the sidelines. So the question of what guardrails are needed in an increasingly digitalizing economy and society have to be raised by corporate whistle blowers like Frances Haugen, who worked at Facebook ("Frances Haugen" 2021).

Digital tools and services have the potential to improve our lives even more than electricity has enabled them over the past two centuries, but like fire or plastic they come with greater dangers than are visible at first use (Fung & Meinel, 2021). The sensors of the Internet of Things (IOT) and 5G are beginning to enable precise tracking of movement, temperature and humidity. This can eventually lead to enhanced food security and efficiencies in agriculture, transportation and the built environment, truly enhancing human wellbeing and our ability to flourish while achieving this with less impact on the earth's resources. As a tool to measure and monitor climate and the environment, digitally enabled sensing can enable us to become better stewards of our planet at a time when stewardship is desperately needed.

3   FALSE CHOICES AND REAL CHOICES

This moment, so full of both positive potential and danger, requires our full attention and consideration.

One set of choices is whether we go down the road of American Capitalist Surveillance (where the likes of Amazon, Google and Facebook have collected so much data about us that they can feed our preferences and steer our behaviour through advertising) or Chinese State Surveillance (which, on the one hand, can usefully curtail fake science, but which also curates access to news). There is also a third option: to backtrack or put on the brakes (which is partly what the Europeans are doing with privacy protection laws in the General Data Protection Regulation (GDPR).

What if this framing is a false choice? What are the underlying assumptions related to judging what is good and meaningful? For whom? False choices arise when we are in the grip of old rituals and fears. Real choices emerge when we allow ourselves to be curious, to inquire and imagine together. Moral imagination is that ability of human beings to use thinking, feeling and reflection, grounded in lived experiences, to then articulate what it is that WE really want.

What do we want? We want a good life for ourselves and our children, even for our neighbours, our politicians and our leaders, and for those we buy from and sell to, for those who grow our food, weave our clothes and cobble our shoes. But in a time of transition such as we are now experiencing, it is a false path to set up our choices as binary, as this or that, us or them.

For example, in the initial stages of the Covid-19 pandemic, the binary choice set up was between individual rights versus collective good. So the world split into two camps, one side (mostly Western) alarmed at infringement of individual rights, and the other side (mostly Asian, Latin American, African) alarmed at the cavalier disregard for the collective good.

The personal or the collective? Fons Trompenaars (Trompenaars & Hampden-Turner 1997), an early thinker in examining diversity from the perspective of mental models, has pointed out that dichotomous thinking pits one form of the good against another form of the good. We cannot neglect the collective even as we cannot deny the personal. We are in a dilemma, but the way forward should not be about choosing one over the other, but finding a way to satisfy both.

4   To Resolve the Dilemma of False Choices: Reconciliation

In his workshops, Trompenaars uses the example of a cylinder: viewed from the top it is a circle, viewed from the side it is a square or rectangle. The cylinder reconciles or integrates both circle and square. This is not a compromise or a balance; it is not 50% a circle or 50% a square. It is fully both, and more.

So how can we balance individual and collective needs? Societies with adequate wealth and public health science have done well, both in the West (Scandinavia, Germany, New Zealand) and the East (Singapore, Japan, South Korea). In these countries, the deaths per million are among the lowest in the world. At the same time, the collective social fabric has been left intact, a difficult balance. In China, the state, by imposing severe lockdowns, suppressed individual rights. Yet, once collective health was secured, personal freedoms were quickly returned, with the Chinese enjoying normal lives at school, work and play through the past two years of the pandemic. Europe has taken the lead in privacy protections with the GDPR (General Data Protection Regulation 2021) which emphasizes the right to personal privacy, but GDPR is today making it difficult to conduct scientific research into health—requiring rethinking of the approaches to data sharing for scientific purposes. India has recently announced the Account Aggregator (Singh, 2021) in order to make it easier for small and medium-sized businesses to access financing in a country where many people and many businesses do not have bank accounts.

There are many explorations underway as every country in the world is working out its own path forward towards reconciling personal and collective good. All are experimenting, flexing, adapting and adjusting as needed (Ritchie et al., 2020). Current science and today's university disciplines are not set up to consider these new explorations in ways that can allow us to share and learn together. Early on, sciences did not have laboratories, but once there was an standardized laboratory environment for experimenting, many more people could contribute their knowledge by conducting laboratory experiments and sharing results.

Today, what we lack in advancing digital humanism is a network of Community Living Learning Labs, where we can compare these disparate digital experiments underway around the world. And we need to conduct these experiments not at the scale of the 100's of millions of people in the EU, USA, China or India, but with the ability to look, as under a microscope, at the interactions between people and how cumulatively these interactions affect the whole society around them, and the local, regional, national and global economy.

To conclude this section, we list certain other polarities that more usefully viewed as dilemmas, not binary wars.

- Economic growth / ecological sustainability
- Rewards for individual economic risk taking/lessening the pain of economic restructuring
- Surveillance to enhance social order/privacy to enhance personal subjectivity
- Profitable/equitable.
- Uniformity leading to scale/diversity leading to localization
- Freedom/responsibility

The digital dimension is present in all these dilemmas, as an accelerator, disruptor, solution or irritant. In itself, digital development is not going to lead to resolution. We, the people, will need higher-order ways of thinking in order to develop new perspectives and reframe these false choices.

5 THE DIGITAL TSUNAMI—AN ARCHIPELAGO'S HISTORICAL JOURNEY INFORMS OUR DIGITAL AGE

Nusantara (*nusa* = island, *antara* = between or among) is the Indonesian word for archipelago, and the world's largest archipelago is located in the South China Sea. It comprises the 17,000 + islands of Indonesia, the 7,600 + islands of the Philippines, and the island of Singapore, where both of us were born. The spread of Southeast Asian islands from Sumatra in the west to Timor Leste in the east equals the distance from Seattle to Cuba, or from London to Afghanistan. The islands are separated not just by seas, but also by language: Seven hundred languages in Indonesia are spoken by 1300 ethnic groups; The Philippines has 175 ethnolinguistic groups (Ethnic groups in the Philippines, 2021); Myanmar has 135 officially recognized ethnic groups (List of ethnic groups in Myanmar, 2021), as well as more unofficial, unrecognized ones like the Rohingya. Southeast Asia has 1000 + languages within a population of 650 million people. Is there something for us to learn from the history of this archipelago, whose diverse peoples have learned to live together—separated in language and islands, but engaging together in trade and interchange?

Spices were crucial for medieval Europe as sources of preservatives, medicine and luxury. When the Ottoman Empire cut off the land routes, Europeans searched for a sea route to India, which they thought was the source of spices. By following the routes of Indian and Arab traders, they then discovered Southeast Asia. With European's advanced military science—guns and boats capable of projecting power across distances – and by lending this power to disputing local rulers, these outsiders gained control that eventually grew into colonization.

The people of the Maluku Islands (Moluccas) did not understand the danger they were in when they were the only source of nutmeg (Maluku Islands, 2021), a spice considered crucial for the treatment of the bubonic plague in Europe (Kiefaber, 2016). Instead, following the trading ethos of their time, the Moluccans sold to all Europeans, just as they had previously sold to all Arabs and Chinese. However, with fierce rivalry among the European powers back home, the Dutch sought to monopolize the nutmeg trade.

The native Moluccans, who had traded nutmeg with Arabs, Indians and Chinese for centuries, were mystified by the Dutch. They did not understand the idea of monopoly. They also did not understand how the Dutch would use their inter-group rivalry against them. The Moluccans remained ignorant of their own true worth and what was underway in the wider world of trade. Their local leaders were focused on seeking ascent to the throne and they neglected to consider the long-term prospects for their people.

The colonizers were able to gain control by subjugating or killing the locals and bringing in workers from other places to grow and harvest the nutmeg trees. To keep prices high, nutmeg was eliminated everywhere else it grew, bumper crops were burned and seeds were soaked in lime to kill them. We are seeing history repeat itself today.

We have emerging digital colonizers: Alphabet, Amazon, Apple, Meta (Facebook), Microsoft and Netflix are now 23% of the Standard and Poors Index, with valuations from $1 trillion to $2 trillion (Big Tech 2021). With an emphasis on getting quickly to global scale they bring a one-size-fits-all mindset even as more and more countries in the world are adopting their platforms and using their services. And this is not by their force, it's by our choice. In October 2021, US "big tech" was compared to "big tobacco" for prioritizing profits above people's health and safety (Kamp, 2021). In China, Baidu AI, Alibaba Cloud, TenCent Cloud and Huawei Cloud have come under increasing regulatory scrutiny. As Kate Park notes, "Beijing passed the Data Security Law in June that started to go into effect in early September for protecting critical data related to national security and issued draft guidelines on regulating the algorithms companies, targeting ByteDance, Alibaba Group, Tencent and DiDi and others, in late August" (Park, 2021).

Colonizers in the twenty-first century are facing a different global environment and scrutiny in this digital age. Meta (Facebook) cannot walk away from whistleblowers forever. The importance with which the latest Facebook whistleblower is being treated is underlined by corporate instructions on how Meta (Facebook) employees should talk to their family members about the allegation (Isaac et al., 2021). An initiative from the United Nations (Gutteres, 2021) produced a roadmap for Digital Cooperation that deserves more attention because it offers a new way forward for us to come together for a better future for humanity.

A safe, inclusive and equitable digital future is essential for progress and peace. The Roadmap for Digital Cooperation that I launched in 2020 offers a vision for a digitally interdependent world that "Connects, Respects and Protects" all people, in which all can thrive, and in which digital tools do not cause harm or reinforce inequalities but instead are a force for good …. Wide-ranging efforts across a year of global upheaval caused by the COVID-19 pandemic have highlighted both opportunities and risks, and underscored the need for stronger collective stewardship.

The colonialists exploited local divisions, as the powerful have always done. In the face of today's seemingly overwhelming technological power, Southeast Asian people, having seen the story play out before, are quietly understanding and taking stock of their own longer-term interests and seeking to pursue them, cognizant of the lessons of history.

The Association of Southeast Asian Nations (ASEAN) is one of the global leaders in regional digital cooperation. The ASEAN digital sector is mindful about the gaps in the level of development across the ASEAN member states. By conducting stocktaking studies across ASEAN member states on specific policy issues and holding workshops to identify and distil best practices, ASEAN member states are advancing their respective national development in the areas of concern guided by regional best practice and national priorities.

Policy areas include broadband infrastructure deployment, bridging the gaps in access to ICT among the population (digital divide), adoption of emerging technologies, radio frequency spectrum management, telecommunication services regulation, intra-ASEAN submarine cable connectivity resilience, network security, and critical information infrastructure protection. The digital sector has also been organizing regular competition-based events to promote the development of digital talents and the growth of digitally-enabled innovation, especially among ASEAN youth and young professionals. Among the annual competition-based events are the ASEAN Cyber-Kids Camp (since 2009), ASEAN ICT Innovation Award (since 2011), ASEAN Makers Hackathon (since 2017) and ASEAN-Japan Cyber SEA Game (since 2018).

In January 2021 ASEAN approved the Data Management Framework (ASEAN Secretariat 2020), which provides a step-by-step guide for businesses and small and medium-sized enterprises (SMEs), to put in place a data management system, which includes data governance structures and safeguards. Good data management practices are key to helping businesses unlock the value of data while ensuring adequate safeguards. This helps reduce the negotiation and compliance cost and time, especially for SMEs, while ensuring personal data protection when data is transferred across borders.

Modern-day Southeast Asian geopolitics is informed by both colonial history and pre-colonial history. Consciousness of this legacy is helpful for the rest of the world facing the prospect of digital colonization. Beyond the US or Chinese examples, there are paths to be pioneered which are equitable and inclusive respecting diverse cultures, generations, and perspectives.

ASEAN has maintained peace in a region that is the most diverse in the world, including being the country with the largest population of Muslims in the world (Indonesia), the country with the third-largest Catholic population (the Philippines), the country with the second-largest number of Buddhists (Thailand) and the world's second-largest communist country (Vietnam).

Europe was rent apart through the ages by religious and ideological wars (Buc, 2015), and in the early part of the twentieth century by nationalistic rivalries that resulted in two world wars. The EU was constituted to prevent such wars and is a supra-national entity with the power to impose its will on member states. When it was formed, the national sensitivities of its members, long a point of conflict, had to be addressed. The decision was made that, in all proceedings, documents must be translated into 22 languages (Mahbubani & Sng, 2017): universalism is achieved through particularism.

New models of political cooperation are emerging in the Southeast Asian region, which has had more bombs dropped on it than Europe had in WW2 (World War II). ASEAN has preserved the peace by showing public respect for each other and by bringing leaders together over and over again. Golf games are a staple of the ASEAN diplomatic trust-building process. Yearly summits include doing skits, at which visitors from the US or EU are awkwardly compelled to perform. One of us went to an ASEAN meeting where the host, the Philippines Foreign Minister, took to the microphone and sang a song to welcome delegates. That broke the ice.

ASEAN governments are democratic, communist and socialist, and they include sultanates, monarchies and military juntas. When the Philippines, a country which has more Catholics than the USA or Italy, hosted

ASEAN leaders in 2018, it chose to stage the Hindu epic *Ramayana* to welcome its neighbours. Why would a Catholic country do that? It understood that the region had been shaped by Indic history, languages and culture, and paid homage to that.

ASEAN has been criticized for being a talk-shop, ineffective in speaking out against members who abuse human rights. But these are Western expectations. In the ASEAN local contexts, an association allows very different types of neighbours to adjust and find ways to get along, respecting national sovereignty. In the face of the social and economic disruption of the Covid pandemic, could attention to mutual respect help devise regional cooperation that works better for the people and the planet?

The Asia Foundation worked with Google.org on a digital literacy initiative, "Go Digital ASEAN", to train up to 200,000 people in marginalized communities across ten countries together with local non-profit companies. By taking a proactive, relationship-building approach, ASEAN countries are embracing the big tech companies (Google.org is part of Alphabet) and working to increase digital inclusion and financial literacy so that the region's small and medium businesses can be economically viable and sustainable. Support is both short term—grants to help vulnerable communities get through the Covid pandemic—and long term—grants to bring more people into the digital ecosystem by training students in coding, with an emphasis on girls, young people and people from vulnerable communities.

6   BUILDING TRUST—A HISTORICAL JOURNEY

Trust is the ubiquitous currency for all human relations, regardless of culture. Yet, culture determines how one builds trust. The EU uses a legal framework with explicit procedures and rules, and sanctions for departing from them. ASEAN cohesion consists of making leaders and bureaucrats comfortable with each other through golf games and skits. Going to karaoke clubs is a large part of doing business in Asia. In Asia, where hierarchy and formality are important aspects of relationship, we think that there must be a way to also experience equality, vulnerability and humanity, and karaoke is a fun way to do so. The Canadian-born philosopher Daniel Bell writes, in *China's New Confucianism*, that karaoke is a ritual, *li*, which helps ease relationships and foster closeness (Karaoke, Ancient Harmony? 2011).

Mahbubani and Sng (2017) credit this low-key, high-touch approach to the Indonesian/Javanese concepts of *musyawarah* and *mufakat,* or consultation and consensus. A vote is hardly ever taken; instead, conversations build upon conversations. European cultures, particularly in the Protestant north, have been described as low-context cultures in which explicit rules guide behaviour. Asia is the opposite. Context matters. And so, tremendous efforts are made to foster contexts that encourage building trust.

In Southeast Asia, self-understanding has been shaped largely by how the West has defined the non-Western. The West, on the other hand, understands itself based on its own self-description. A switch—the West listening to how others think of it, and the non-Western world trying to understand itself on its own terms—is happening, and this is necessary for a new humanism. This allows for a dual process: becoming aware and critical of how others understand us, and coming to understand ourselves on our own terms.

The foremost historian of Asia, Prof Wang Gung Wu, OBE, AO, and university professor at the National University of Singapore (NUS), tells the story of Southeast Asia through multiple lenses provided by different streams of historical scholarship. Shaped by the demands of geography and incessant war, the Chinese Empire centralized 2200 years ago and has ever since pursued stability as a goal—fending off land attacks from the north and west. The West, founded in the Greco-Roman Mediterranean, has been both a land and sea power, rivalrous and fragmented. Due to military sophistication, a result of ongoing rivalry among its different members, the West has learned to project power far and wide.

Southeast Asian empires in the Indonesian archipelago took a different turn. Inhabited 1.5 million years ago by Java Man, the later history of the region was influenced by trade and religion coming from India. We mention prehistory because it shows that the climate here was hospitable to humanoids for a far longer time than the Mediterranean. People evolved a cultural rhythm which embraces diversity. One of the earliest empires was the 7th-century naval kingdom of Srivijaya, which brought Hinduism and Buddhism together; a testimony to this is the world's largest Buddhist temple, built at Borobudur, Indonesia. The 13th-century Hindu Majapahit Kingdom still has a strong legacy on present-day Bali, Indonesia.

The peaceful spread of Islam in Southeast Asia occurred through trade and not via conquest. Muslims have related to others, and others have connected with Islam here in historical ways that are neither fraught nor traumatic, and Islam is a religion that historically has served as a stabilizer (despite the radicalism of the last few years that has been exported here from the Middle East) This layering of centuries of civilization, like a *kueh lapis* (a nine-layered cake), is true of other parts of Southeast Asia, including a Sinic layering of Confucianism and Taoism. Only after the arrival of the Europeans and the advent of Christianity and modernity did the later layering become a very violent one. There were battles prior to the Europeans, but the paradigm for prosperity and peace in Southeast Asian history was mutually beneficial trade, not conquest.

Silk and porcelain from China, textiles and medicines from India, and spices from the region circulated to east, west, north and south. The civilizing wisdom of Islam, Hinduism, Buddhism, Taoism and Confucianism also moved and intermingled through trade. The result was mixed marriages and mixed cuisines. It is not uncommon throughout Southeast Asia to find mosques, temples and churches built alongside each other, as testimony not only to tolerance, but to mutual appreciation.

Clashes have included race riots and religious conflicts (Singapore, Malaysia and Indonesia with domestic race riots in the 1960s, Indonesia in the 1990s, Cambodia versus Vietnam in the 1970s, Cambodia versus Thailand in the 2000s, the Rohingya in Myanmar). When conflict origins are indigenous, their severity and longevity have generally been exacerbated by outsiders, in particular the great European powers historically, and Meta (Facebook) in the more recent case of the Rohingya. With the deep religiosity and ethnic ties of the region, peace prevails through the rituals of material and social intercourse. In Surabaya recently, a group of religious leaders gathered to build a Buddhist temple, Confucian temple, Protestant church and Catholic church adjoining a mosque (Channel News Asia 2020).

Mabubhani and Sng argue that ASEAN's neighbourliness—the golf games, skits, songs, trade, consultations, consensus and non-interference—have preserved peace and prosperity. ASEAN and Southeast Asia can be characterized by a sensibility: seek trade; find mutual benefits; work hard; chill out.

Art, literature, film, music, visual culture, media, and animation, gaming, and Internet culture are transforming the region as digital media changes the world of politics, economies, social lives and culture. Traditional cultural and scholarly works are being digitally transformed. Virtual/augmented reality digital communities are inventing new forms of cultural interactions, which in turn are intervening and reshaping the non-virtual reality. The emergence of digital cultural production is changing art and communication, with ASEAN digital culture introducing new perspectives in the connected world and showing how personal and social lives and societal institutions and rules are being mediated through digital representation. Research in humanities must evolve in the digital age, new academic disciplines must be recognized. We need a higher-order discussion about what is needed for humans when technology affects so many parts of living, culture, work and play. It is not surprising that in the face of all this change, people are fearful, frustrated and yes, behaving foolishly when the guardrails are not in place. Digital humanism offers a broad umbrella to both preserve the different stories from different traditions and bring us together to build new narratives (and guardrails) for our global digital age.

7   Digital Humanism: No Longer History But Our Story

One of us was a close colleague of Doug Engelbart, the inventor of the computer mouse and, as the father of the human–computer interface, one of the founding fathers of digital technology. Engelbart warned against the *personal* computer, believing that such technological power should not be for making individuals powerful, but for enabling new methods of sharing and learning in networks of communities. Nevertheless, his team at SRI left him, abandoned him by his telling, to go to Xerox PARC (Markoff, 2005) to develop the personal computer. The vision of digital technology serving living learning communities connected in networks had to wait for a different day, and perhaps that time is now.

Digital development, like much else, has been shaped by the"hero's journey" (Hero's journey, 2021) and the heroic stories of mythical and historical figures like Achilles, Odysseus, Prometheus, John Galt, Vasco Da Gama, Christopher Columbus, Cortez, Johan van Oldenbarnevelt, Sir Walter Raleigh and Sir Stamford Raffles. Corporate titans Steve Jobs, Jeff Bezos, Elon Musk and Mark Zuckerberg are just the latest in this archetypal motif. Their stories focus on pioneering, exploring, taking risks, and capturing and perpetuating individual wealth. This has unleashed tremendous energies and also created unprecedented negative consequences.

In the Asian context, the hero that has captured the imagination is someone who triumphs over self rather than over external enemies. Professor Dennis Kratz, a Greek medievalist who is also director of the Center for Asian Studies at The University of Texas at Dallas, compares the Monkey King (SunWuKung) to Odysseus. The latter fears the mortality and meaninglessness of a human life that achieves nothing, and therefore he strives for glory and achievement. The Monkey King, the Chinese version of the Hindu deity Hanuman, is an allegory of the human being, with a choice to make: to be chained to the desires of the egoic self, or to be liberated.

As portrayed in the Ming dynasty novel, *Journey to the West, the Monkey King*, which is well known to children all over East and Southeast Asia, he is a figure with awesome magical powers for good or evil. In the Ming re-telling of an actual historical event during the Tang Dynasty, he is pressed into helping the monk Xuanzang on his journey to India to obtain Buddhist scriptures.

The Monkey King, who is a wild demonic figure, reaches India and becomes an enlightened Arhat, which is the first stage of Buddhahood. In Kratz's analysis, this amalgamation of Hindu-Taoist-Buddhist-Confucian beliefs says that the human ego, left to its own delusions of greed and hatred, of cravings and aversions, is an animal and a demon, although also powerful. But the self, when cultivated and disciplined (Confucian teachings) can do good. And when it sees into the ephemeral nature of the cosmos and realizes that it too is impermanent (Buddhist teachings) it becomes an awakened Buddha (Kratz, 2020). Moreover, in Taoist fashion, all the animal and plant spirits portrayed in the novel are likewise capable of becoming Buddhas. This is an egalitarian and optimistic view of the cosmos.

As we think about the awesome powers unleashed by digital development, as well as the economic wealth it can create, the issue of how to be human must be re-examined. There is, on the one hand, the "heroic conqueror" mode that sees the digital revolution as bringing Promethean fire. Yet we can, at the same time, also ask questions about what is really hurting and harming us. What is causing the psychic pain we are collectively suffering? We can also aspire to evolve. There is the possibility to learn to master our demonic, lustful, greedy, out-of-control natures, and to awaken to the purpose of liberating other sentient beings from suffering. To leave a better world for future generations who are being harmed by our choices taken today without proper consideration of the monumental effect on their future lives.

Digital literacy is more than just learning the rules of the digital road, it is more than learning to code, it is admitting our lack of understanding and awareness about how profoundly, intimately and collectively we have allowed digital technology tools, assumptions and services to pervade our lives, our families and our communities. We should respond appropriately to this mighty challenge.

8   Conclusion

The digital world is the new dimension. Becoming a different kind of human being is also the next dimension. If our humanism – our humanity itself – is not developed further and doesn't evolve, then our digital technology becomes a set of demonic Promethean powers we have unleashed but cannot master.

The quintessential question our grandmothers asked us is this:

你怎么做人啊? How do you want *to do* human?

We must speak up for our essential natures and cultures, and harness and evolve compassion and care. The platforms are there for voices to be heard and new alliances to be made. Digital humanism allows and requires this, even if the Tech Masters of the Universe do not want it. People must actually think differently, venture capital must fund differently and entrepreneurs must heed Abraham Lincoln's call:

> We can succeed only by concert. It is not "can *any* of us *imagine* better?" but "can we *all* do better?" The dogmas of the quiet past, are inadequate to the stormy present. The occasion is piled high with difficulty, and we must rise—with the occasion. As our case is new, so we must think anew, and act anew. We must disenthrall ourselves, and then we shall save our country. (Lincoln, 1989, pp. 414–415)

References

Association of Southeast Asian Nations (ASEAN) Secretariat. (2020). *ASEAN Digital Sector: Cooperation activities*. Retrieved October 13, 2021 from

https://asean.org/our-communities/economic-community/asean-digital-sec
    tor/
Big Tech. (2021, October 12). *Wikipedia*. Retrieved October 13, 2021 from
    https://en.wikipedia.org/wiki/Big_Tech
Buc, P. (2015). *Holy war, Martyrdom, and terror: Christianity, violence, and the
    West*. University of Pennsylvania Press.
Ethnic groups in the Philippines. (2021, 23 September). *Wikipedia*. Retrieved
    October 13, 2021 from https://en.wikipedia.org/wiki/Ethnic_groups_in_
    the_Philippines
Frances Haugen. (2021, 12 October). *Wikipedia*. Retrieved 17 October 2021
    from https://en.wikipedia.org/wiki/Frances_Haugen
Fung, M. L., & Meinel, C. (2021, October 11). *Clean IT poli-
    cies to support sustainable digital technologies*. Task Force 7 Climate
    Change, Sustainable Energy & Environment. Retrieved October 18,
    2021 from https://www.g20-insights.org/policy_briefs/clean-it-policies-to-
    support-sustainable-digital-technologies/
General Data Protection Regulation. (2018). *Intersoft consulting*. Retrieved 21
    October 2021 from https://gdpr-info.eu/
Gutteres, A. (2021). *Roadmap for digital cooperation. United Nations Office of
    the Secretary General*. Retrieved October 13 2021 from https://www.un.
    org/techenvoy/sites/www.un.org.techenvoy/files/Update_on_Roadmap_i
    mplementation_April_2021.pdf
Hero's journey. (2021, October 1). *Wikipedia*. Retrieved October 13, 2021 from
    https://en.wikipedia.org/wiki/Hero%27s_journey
Isaac,M., Mac, R., & Frenkel, S. (2021, October 10). After whistleblower goes
    public, Facebook tries calming employees. *The New York Times*. Retrieved
    October 13, 2021 from https://www.nytimes.com/2021/10/10/techno
    logy/facebook-whistleblower-employees.html
Kamp, L. (2021, October 11). Lena Kamp: Facebook prioritizing profit over
    security. *Humanity in Action*. Retrieved October 13, 2021 from https://
    www.humanityinaction.org/news_item/news-item-germany-lena-kampf-fac
    ebook-prioritizing-profit-over-security/
Karaoke, ancient harmony?. (2011). *The Diplomat*. Retrieved October 13, 2021
    from https://thediplomat.com/2011/04/karaoke-ancient-harmony/
Kiefaber, D. (2016). The darker, spicier history of nutmeg. *InsideHook*.
    Retrieved October 13, 2021 from https://www.insidehook.com/article/
    food-and-drink/spicy-history-nutmeg
Kratz, D. (2020, September 29). *Odysseus meets the Monkey King: Journeying
    West, East, and beyond*. (Video) YouTube. https://www.youtube.com/watch?
    v=n7ufkZU7Gmw

Kristof, N., & Dunn, W. D. (2020). Who killed the Knapp family?. *The New York Times*. Retrieved October 13, 2021 from https://www.nytimes.com/2020/01/09/opinion/sunday/deaths-despair-poverty.html

Lincoln, A. (1862). *Abraham Lincoln: Speeches and writings 1859–1865* (ed.1989). Library of America.

List of ethnic groups in Myanmar. (2021, 28 July). *Wikipedia*. Retrieved October 13, 2021 from https://en.wikipedia.org/wiki/List_of_ethnic_groups_in_Myanmar

Mahbubani, K., & Sng, J. (2017). *Chapter 6*. National University of Singapore Press.

Maluku Islands. (2021, 6 October). *Wikipedia*. Retrieved October 13, 2021 from https://en.wikipedia.org/wiki/Maluku_Islands

Markoff, J. (2005). *What the Dormouse said: How the sixties counterculture shaped the personal computer industry*. Viking Press

Park, K. (2021). Chinese crackdown on tech giants threatens its cloud market growth. *Techcrunch*. Retrieved October 13, 2021 from https://techcrunch.com/2021/09/13/chinese-crackdown-on-tech-giants-threatens-its-cloud-market-growth/

Ritchie, H., Mathieu, E., Rodés-Guirao, L., Appel, C., Giattino, C., Ortiz-Ospina, E., Hasell, J., Macdonald, B., Beltekian, D., & Roser, M,. (2020). Coronavirus pandemic (Covid 19). *Our World in Data*. Retrieved October 13, 2021 from https://ourworldindata.org/policy-responses-covid

Singh, M. (2021, September 2). India launches Account Aggregator system to extend financial service to millions. *Techcrunch*. Retrieved October 13, 2021 from https://indialaunches-account-aggregator-system-to-extend-financial-services-to-millions

Trompenaars, F., & Hampden-Turner, C. (1997). *Riding the Waves of Culture: Understanding Cultural Diversity in Business*. Retrieved October 13, 2021from https://ocan.yasar.edu.tr/wp-content/uploads/2013/09/Riding-the-waves_Part-1.pdf

Wang, Y., & Kosinski, M. (2018). Deep neural networks are more accurate than humans at detecting sexual orientation from facial images. *Journal of Personality and Social Psychology, 114*(2), 246–257. Retrieved October 13, 2021 from https://doi.org/10.1037/pspa0000098

Young, D. (Executive producer). (2020). *The mark of empire*. Channel News Asia.

Index